U.S.NRC

United States Nuclear Regulatory Commission

Protecting People and the Environment

NUREG-1805
Supplement 1, Vol. 1

I0482732

Fire Dynamics Tools (FDT^S) Quantitative Fire Hazard Analysis Methods for the U.S. Nuclear Regulatory Commission Fire Protection Inspection Program

Supplement 1

Office of Nuclear Regulatory Research

AVAILABILITY OF REFERENCE MATERIALS
IN NRC PUBLICATIONS

NRC Reference Material

As of November 1999, you may electronically access NUREG-series publications and other NRC records at NRC's Public Electronic Reading Room at http://www.nrc.gov/reading-rm.html. Publicly released records include, to name a few, NUREG-series publications; *Federal Register* notices; applicant, licensee, and vendor documents and correspondence; NRC correspondence and internal memoranda; bulletins and information notices; inspection and investigative reports; licensee event reports; and Commission papers and their attachments.

NRC publications in the NUREG series, NRC regulations, and Title 10, "Energy," in the *Code of Federal Regulations* may also be purchased from one of these two sources.
1. The Superintendent of Documents
    U.S. Government Printing Office
    Mail Stop SSOP
    Washington, DC 20402–0001
    Internet: bookstore.gpo.gov
    Telephone: 202-512-1800
    Fax: 202-512-2250
2. The National Technical Information Service
    Springfield, VA 22161–0002
    www.ntis.gov
    1–800–553–6847 or, locally, 703–605–6000

A single copy of each NRC draft report for comment is available free, to the extent of supply, upon written request as follows:
Address: U.S. Nuclear Regulatory Commission
         Office of Administration
         Publications Branch
         Washington, DC 20555-0001
E-mail: DISTRIBUTION.RESOURCE@NRC.GOV
Facsimile: 301–415–2289

Some publications in the NUREG series that are posted at NRC's Web site address http://www.nrc.gov/reading-rm/doc-collections/nuregs are updated periodically and may differ from the last printed version. Although references to material found on a Web site bear the date the material was accessed, the material available on the date cited may subsequently be removed from the site.

Non-NRC Reference Material

Documents available from public and special technical libraries include all open literature items, such as books, journal articles, transactions, *Federal Register* notices, Federal and State legislation, and congressional reports. Such documents as theses, dissertations, foreign reports and translations, and non-NRC conference proceedings may be purchased from their sponsoring organization.

Copies of industry codes and standards used in a substantive manner in the NRC regulatory process are maintained at—
> The NRC Technical Library
> Two White Flint North
> 11545 Rockville Pike
> Rockville, MD 20852–2738

These standards are available in the library for reference use by the public. Codes and standards are usually copyrighted and may be purchased from the originating organization or, if they are American National Standards, from—
> American National Standards Institute
> 11 West 42nd Street
> New York, NY 10036–8002
> www.ansi.org
> 212–642–4900

United States Nuclear Regulatory Commission

Protecting People and the Environment

NUREG-1805
Supplement 1, Vol. 1

Fire Dynamics Tools (FDT^s) Quantitative Fire Hazard Analysis Methods for the U.S. Nuclear Regulatory Commission Fire Protection Inspection Program

Supplement 1

Manuscript Completed: June 2013
Date Published: July 2013

Prepared by:
D. Stroup*, G. Taylor*, G. Hausman**

*Office of Nuclear Regulatory Research
**Region III

M. H. Salley, NRC Project Manager

Office of Nuclear Regulatory Research

ABSTRACT

The U.S. Nuclear Regulatory Commission (NRC) has developed quantitative methods, known as "Fire Dynamics Tools" (FDT^s), for analyzing the impact of fire and fire protection systems in nuclear power plants (NPPs). These methods have been implemented in spreadsheets and taught at the NRC's quarterly regional inspector workshops. The FDT^s were developed using state-of-the-art fire dynamics equations and correlations that were preprogrammed and locked into Microsoft Excel® spreadsheets. These FDT^s enable inspectors to perform quick, easy, first-order calculations for potential fire scenarios using today's state-of-the-art principles of fire dynamics. Each FDT^s spreadsheet also contains a list of the physical and thermal properties of the materials commonly encountered in NPPs.

This NUREG-series report documents a new spreadsheet that has been added to the FDT^s suite and describes updates, corrections, and improvements to the existing spreadsheets. The majority of the original FDT^s were developed using principles and information from the Society of Fire Protection Engineers (SFPE) *Handbook of Fire Protection Engineering*, the National Fire Protection Association (NFPA) *Fire Protection Handbook*, and other fire science literature. The new spreadsheet predicts the behavior of power cables, instrument cables, and control cables during a fire. The thermally-induced electrical failure (THIEF) model was developed by the National Institute of Standards and Technology (NIST) as part of the Cable Response to Live Fire (CAROLFIRE) program sponsored by the NRC. The experiments for CAROLFIRE were conducted at Sandia National Laboratories, Albuquerque, New Mexico. THIEF model predictions have been compared to experimental measurements of instrumented cables in a variety of configurations, and the results indicate that the model is an appropriate analysis tool for NPP applications. The accuracy and simplicity of the THIEF model have been shown to be comparable to that of the activation algorithms for various fire protection devices (e.g., sprinklers, heat and smoke detectors).

CONTENTS

FIGURES

TABLES

EXECUTIVE SUMMARY

The U.S. Nuclear Regulatory Commission (NRC) has developed quantitative methods, known as "Fire Dynamics Tools" (FDT^s), for analyzing the impact of fire and fire protection systems in nuclear power plants (NPPs). These methods have been implemented in spreadsheets and taught at the NRC's quarterly regional inspector workshops. The goal of the training is to assist inspectors in calculating the quantitative aspects of a postulated fire and its effects on safe NPP operation. The FDT^s were developed using state-of-the-art fire dynamics equations and correlations that were preprogrammed and locked into Microsoft Excel® spreadsheets. These FDT^s enable inspectors to perform quick, easy, first-order calculations for potential fire scenarios using today's state-of-the-art principles of fire dynamics. Each FDT^s spreadsheet also contains a list of the physical and thermal properties of the materials commonly encountered in NPPs.

The FDT^s are intended to assist fire protection inspectors in performing risk-informed evaluations of credible fires that may cause critical damage to essential safe-shutdown equipment, as required by the reactor oversight process (ROP) defined in the NRC's inspection manual. In the ROP, the NRC is moving toward a more risk-informed, objective, predictable, understandable, and focused regulatory process. Key features of the program are a risk-informed regulatory framework, risk-informed inspections, a significance determination process (SDP) to evaluate inspection findings, performance indicators, a streamlined assessment process, and more clearly defined actions that the NRC will take for plants based on their performance.

This NUREG-series report documents a new spreadsheet that has been added to the FDT^s suite and describes updates, corrections and improvements for the existing spreadsheets. The majority of the original FDT^s were developed from the Society of Fire Protection Engineers (SFPE) *Handbook of Fire Protection Engineering*, the National Fire Protection Association (NFPA) *Fire Protection Handbook*, and other fire science literature. The new spreadsheet predicts the behavior of power cables, instrument cables, and control cables during a fire. The thermally-induced electrical failure (THIEF) model was developed by the National Institute of Standards and Technology (NIST) as part of the Cable Response to Live Fire (CAROLFIRE) program sponsored by the NRC.

The primary objective of CAROLFIRE was to characterize the various modes of electrical failure (e.g. hot shorts, shorts to ground) within bundles of power, control and instrument cables. A secondary objective of the project was to develop a simple model to predict thermally-induced electrical failure (THIEF) when a given interior region of the cable reaches an empirically determined threshold temperature. The experiments for CAROLFIRE were conducted at Sandia National Laboratories.

The THIEF model for cables has been shown to work effectively in realistic fire environments. The THIEF model is essentially a numerical solution of the one dimensional heat conduction equation within a homogenous cylinder with fixed, temperature independent properties. THIEF model predictions have been compared to experimental measurements of instrumented cables in a variety of configurations, and the results indicate that the model is an appropriate analysis tool for NPP applications. The model is of comparable accuracy and simplicity to the activation algorithms for various fire protection devices (e.g., sprinklers, heat and smoke detectors).

ACKNOWLEDGMENTS

Since the publication of NUREG-1805, *Fire Dynamics Tools (FDT^s) Quantitative Fire Hazard Analysis Methods for the U.S. Nuclear Regulatory Commission Fire Protection Inspection Program*, numerous comments and suggestions for additions, improvements, and a few corrections have been received from users throughout the world. This supplement addresses many of the issues identified by users and adds a new spreadsheet to the suite of FDT^s. The authors thank the internal and external stakeholders who have taken the time to provide comments and suggestions on the original report. We hope this supplement will receive similar attention and appreciate any feedback from users of the material in this supplement.

The authors gratefully acknowledge the support and assistance provided by Naeem Iqbal and Mark Henry Salley of the U.S. Nuclear Regulatory Commission (NRC). They published the original NUREG-1805 in December 2004. The general concepts used in creating and developing the FDT^s spreadsheets were similar to those taught by Dr. Frederick Mowrer whose fire modeling course they had attended during their postgraduate studies at the University of Maryland.

We acknowledge and appreciate the contributions of Mollie Semmes, a fire protection engineering student at the University of Maryland. Mollie's hard work and diligence during her summer internships at NRC ensured that this report was published in a timely fashion and with completely revised and tested spreadsheets. The authors also thank Nicolas Melly, David Gennardo, and Kendra Hill in the Fire Research Branch of the NRC Office of Nuclear Regulatory Research for their comments and testing of the spreadsheets.

The new spreadsheet added to the FDT^s implements a methodology for estimating the thermally-induced electrical failure of cables. This THIEF model was derived from an algorithm developed by Dr. Kevin McGrattan at the National Institute of Standards and Technology based on data obtained from cable tests conducted at Sandia National Laboratories by Mr. Steven Nowlen.

Finally, the authors would like to thank H.W. 'Roy' Woods who provided invaluable support in publishing this report. In addition, we greatly appreciate the efforts of Guy Beltz, NRC printing specialist, and Tojuana Fortune-Grasty, NRC NUREG technical editor, whose expertise were critical to ensuring the quality of the published manuscript.

ABBREVIATIONS

ABS	Acrylonitrile Butadiene Styrene
ACRS	Advisory Committee on Reactor Safeguards (NRC)
ADAMS	Agencywide Documents Access and Management System (NRC)
ADS	Automatic Depressurization System
AFFF	Aqueous Film Forming Foam
AFT	Adiabatic Flame Temperature
AFW	Auxiliary Feedwater
AGA	American Gas Association
AHJ	Authority Having Jurisdiction
AISI	American Iron and Steel Institute
AL	Administrative Letter
ALC	Approximate Lethal Concentration
ANS	American Nuclear Society
ANSI	American National Standards Institute
API	American Petroleum Institute
ASCE	American Society of Civil Engineers
ASCOS	Analysis of Smoke Control Systems
ASET	Available Safe Egress Time
ASHRAE	American Society of Heating, Refrigeration, and Air Conditioning Engineers
ASME	American Society of Mechanical Engineers
ASMET	Atria Smoke Management Engineering Tools
ASTM	American Society for Testing and Materials
AT	Auxiliary Transformer
ATF&E	Alcohol, Tobacco, Firearms, and Explosives
AWG	American Wire Gauge
BFC	Bromochlorodifluoro-methane
BFNP	Browns Ferry Nuclear Power Plant
BFRL	Building and Fire Research Laboratory
BL	Bulletin
BLEVE	Boiling Liquid, Expanding Vapor Explosion
BOCA	Building Officials & Code Administration International
BREAK1	Berkeley Algorithm for Breaking Window Glass in a Compartment Fire
BS	British Standard
BTP	Branch Technical Position
BTU	British Thermal Unit
BWR	Boiling-Water Reactor
CAROLFIRE	Cable Response to Live Fire
CCW	Component Cooling Water
CFAST	Consolidate Model of Fire Growth and Smoke Transport
CFD	Computational Fluid Dynamics
CFI	Certified Fire Inspector
CFO	Chief Financial Officer (NRC)
CFR	*Code of Federal Regulations*
CHF	Critical Heat Flux
CIB	Conseil Internationale du Batiment
CIBSI	Chartered Institution of Building Services Engineers
CIO	Chief Information Officer (NRC)
CL.S.PE	Chlorosulfonated Polyethylene

FRXPE	Fire-Retardant Crosslinked Polyethylene
FSSD	Post-Fire Safe-Shutdown
FTA	Federal Transit Authorization
FTMS	Federal Test Method Standard
GDC	General Design Criteria
GL	Generic Letter
GSA	General Service Administration
GSI	Generic Safety Issue
H2O	Water
HBr	Hydrogen Bromide
HCl	Hydrogen Chloride
HCN	Hydrogen Cyanide
HEPA	High-Efficiency Particulate Air Filter
HF	Hydrogen Fluoride
HPCI	High Pressure Cooling Injection
HRR	Heat Release Rate
HTGR	High-Temperature Gas-Cooled Reactor
HVAC	Heating, Ventilation, and Air Conditioning
IAFSS	International Association of Fire Safety Science
IBC	International Building Code
ICBO	International Conference of Building Officials
ICS	Integrated Control System
ICSDTS	International Committee for the Study and Development of Tubular Structures
IE	Initiative Events
IEC	International Electrotechnical Commission
IEEE	Institute of Electrical and Electronic Engineers
IN	Information Notice
INEEL	Idaho National Engineering and Environmental Laboratory
IPEEE	Individual Plant Examination of External Events
ISO	International Organization for Standardization
LAVENT	Link Actuation Vents
LC	Lethal Concentration LCL Lethal Concentration Low
LD	Lethal Dose
LDL	Lethal Dose Low
LEL	Lower Explosive Limit
LER	Licensee Event Report
LFL	Lower Flammability Limit
LIFT	Lateral Ignition and Flame Spread (ASTM E 1321 Standard Test Method)
LLNL	Lawrence Livermore National Laboratory
LNG	Liquified Natural Gas
LOC	Limiting Oxidant Concentration
LOCA	Loss-of-Coolant Accident
LPG	Liquid Propane Gas
LWR	Light-Water Reactor
MCC	Motor Control Center
MCR	Main Control Room
MESG	Maximum Experimental Safe Gap
MOV	Motor-Operated Valve
MQH	McCaffrey, Quintiere, and Harkleroad
NBC	National Building Code
NBR	Nitrile

NBS	National Bureau of Standards
NEA	Nuclear Energy Agency
NEI	Nuclear Energy Institute
NEMA	National Electrical Manufacturers Association
NFC	National Fire Code
NFPA	National Fire Protection Association
NIOSH	National Institute of Occupational Safety and Health
NIST	National Institute of Standards and Technology
NO2	Nitrogen Dioxide
NOUN	Notification of Unusual Event
NPP	Nuclear Power Plant
NRC	U.S. Nuclear Regulatory Commission
NRR	Office of Nuclear Reactor Regulation (NRC)
NUREG	NUclear REGulatory Guide
OCIO	Office of Chief Information Officer (NRC)
OL	Operating License
ORNL	Oak Ridge National Laboratory
OSHA	Occupational Safety and Health Administration
OSU	Ohio State University
PASS	Personal Alert Safety System
PC	Polycarbonate
PDA	Primary Disconnect Assembly
PE	Polyethylene
PEF	Polyethylene Fluoride
PES	Polyethersulphone
PFA	Perfluoroalkoxy Branched Polymers
PMMA	Polymethylmethacrylate
PP	Polypropylene
PPE	Polytetrafluoroethylene
PRA	Probabilistic Risk Assessment
PS	Polystyrene
PTEF	Polytetrafluoroethylene (Teflon®)
PU	Polyurethane
PVC	Polyvinylchloride
PVF	Polyvinylfluoride
RCP	Reactor Coolant Pump
RES	Office of Nuclear Regulatory Research (NRC)
RG	Regulatory Guide
RHR	Residual Heat Removal
RIS	Regulatory Issue Summary
RMV	Respiratory Minute Volume
ROP	Reactor Oversight Process
RTECS	Registry of the Toxic Effects of Chemical Substance
RTI	Response Time Index
RWFD	Red Wing Fire Department
S/G	Steam Generator
SBC	Standard Building Code
SBCCI	Southern Building Code Congress International
SBDG	Standby Diesel Generator
SBR	Styrene Butadiene Rubber
SCBA	Self-Contained Breathing Apparatus

SDP	Significance Determination Process
SER	Significant Event Report
SFPE	Society of Fire Protection Engineers
SI	System International
SNL	Sandia National Laboratories
SOLAS	Safety of Lives at Sea
SONGS	San Onofre Nuclear Generating Station
SPLB	Plant Systems Branch (NRC)
SRP	Standard Review Plan (NUREG-0800)
SSC	Structure, System, and/or Component
TASEF	Temperature Analysis of Structure Exposed to Fire
TCL	Toxic Concentration Low
TDL	Toxic Dose Low
TFE	Tetrafluoroethylene (Teflon®)
THIEF	Thermally-Induced Electrical Failure
TLC	Toxic Concentration Low
TLV	Threshold Limit Value
TNT	Trinitrotoluene
TP	Thermoplastic
TRP	Thermal Response Parameter
TS	Thermoset
TSC	Technical Support Center
TTC	Time-Temperature Curve
TVA	Tennessee Valley Authority
TVAN	Tennessee Valley Authority Nuclear Program
UBC	Uniform Building Code
UEL	Upper Explosive Limit
UFC	Uniform Fire Code
UFL	Upper Flammability Limit
UL	Underwriters Laboratories
UPS	Uninterruptible Power Supply
USFA	United States Fire Administration
UVCE	Unconfined Vapor Cloud Explosion
V&V	Verification and Validation
VRLA	Valve-Regulated Lead Acid
W/D	Weight-to-Heated Perimeter Ratio
XLPE	Crosslinked Polyethylene
XLPO	Crosslinked Polyolefin

NOMENCLATURE

A_c	Compartment floor area
A_e	Surface of element
A_f	Horizontal burning area of fuel
A_H	Ampere hours
A_s	Cross sectional area
A_T	Area of compartment enclosing surfaces (excluding vent areas)
A_v	Area of ventilation openings
B	Flame spread parameter
C	Gas concentration by volume
C	Thermal capacity
C_i	Specific heat of insulation
C_p	Specific heat
C_s	Specific heat of steel
C_v	Specific heat at constant volume
C_{HF}	Critical heat flux for ignition
D	Diameter
D	Heated parameter
D_{SC}	Scaled distance
E	Emissive power
E	Explosive energy released
F	Configuration or shape factor
F	Fire resistance time
F_{TP}	Flux time product
F_c	Float Current per 100 AH
g	Acceleration of gravity
G	Gas discharge rate
H	Thickness of insulation
h	Heat flux time product index
h_c	Compartment height
h_{eff}	Effective heat transfer coefficient
h_{ig}	Heat transfer coefficient at ignition
h_k	Convective heat transfer coefficient
h_v	Height of ventilation opening
H	Thermal capacity of steel section at ambient
H	Height
H_g	Hydrogen gas generation
h_f	Flame height
$h_{f(wall)}$	Wall flame height
$h_{f(wall,line)}$	Line fire flame height
$h_{f(corner)}$	Corner fire flame height
k	Thermal conductivity
k_i	Thermal conductivity of insulation
$k\rho c$	Thermal inertia
K	Mixing efficiency factor
K	Proportionality constant
L_c	Compartment length
L	Length
LFL	Lower flammability limit

m	mass
m_f	Mass of fuel vapor
m_f	Mass of fuel burned
m_p	Mass concentration of particulate
m	Mass flow rate
m_e	Mass entrainment rate
m_f	Mass flow rate of fuel
m_o	Mass flow rate out of enclosure
m_p	Plume mass flow rate
m''	Mass loss rate per unit area
M_p	Mass of particulates produced
N	Number of cells (batteries)
N	Number of theoretical air changes
P	Pressure
q''	Heat flux
q''_{crit}	Critical heat flux
q''_e	External heat flux
q''_{min}	Minimum heat flux required for ignition
q''_r	Radiative heat flux
Q	Volume of air
Q_{total}	Total energy release
Q	Heat release rate or energy release rate
Q_c	Convective energy release rate
Q_{FO}	Energy release rate to cause flashover
Q_{fs}	Full-scale energy release rate
Q_{bs}	Bench-scale energy release rate
R	Radius
R	Radial distance
R	Fire Resistance
RTI	Response time index
S	Visibility
T	Time
T_b	Burning duration
t_D	Detection time
t_{ig}	Ignition time
t_p	Thermal penetration time
t_r	Detector response time
t_t	Smoke transit time
$t_{activation}$	Sprinkler activation time
T	Temperature
T_a	Ambient temperature
T_f	Fire temperature
$T_{FO(max)}$	Post-flashover compartment temperature
T_g	Gas temperature
T_s	Steel temperature
T_{jet}	Ceiling jet temperature
$T_{p(centerline)}$	Plume centerline temperature
$T_{activation}$	Activation temperature
U_{jet}	Ceiling jet velocity
U_w	Wind velocity
U_o	Gas velocity

u^*	Nondimensional wind velocity
V	Volume
V_{def}	Volume of gas for deflagration
W	Fuel exposed width
W_c	Compartment width
W	Weight of steel column per linear foot
W_{TNT}	Weight of TNT
Y_p	Particulate yield
Z	Height of smoke layer interface above floor
Z_o	Hypothetical virtual origin of fire source
Z_p	Fireball flame height
ΔH_c	Heat of Combustion
$\Delta Hc_{,eff}$	Effective heat of combustion
Δt	Time step
ΔT_g	Gas temperature above ambient
ΔT_{ig}	Ignition temperature above ambient
α	Heat transfer coefficient for steel I
α	Yield (fraction of available energy participating in blast wave generation)
α_m	Specific extinction coefficient
χ_r	Fraction of total energy radiated
δ	Thickness
ε	Flame emissivity
Ω	Ventilation factor
θ	Flame title or angle of deflection
ρ	Density
ρ_a	Density of Ambient Air
ρ_c	Density of combustion products
ρ_c	Density of concrete
ρ_F	Density of fuel vapor
ρ_g	Density of gas
ρ_i	Density of insulation
σ	Stefan-Boltzmann constant
τ_o	Detector time constant
ν	Regression rate

Subscripts

a	Ambient
bs	Bench-scale
c	Compartment
c	Combustion
c	Concrete
c	Current
D	Detection
def	Deflagration
e	Convective
e	External
eff	Effective
e	Entrainment

f	Fire
f	Flame
f	Fuel
f(corner)	Corner flame
f(wall)	Wall flame
f(wall,line)	Line fire flame
FO	Flashover
fs	Full-scale
g	Gas
H	Hours
I	Insulation
ig	Ignition
jet	Ceiling jet
m	Extinction
min	Minimum
o	Out
p	Specific
p	Particulate
p	Plume
p	Penetration
r	Radiative
r	Response
SC	Scale
s	Steel
T	Total
Total	Total
t	Transient
TNT	Trinitrotoluene
v	Vent
v	Volume
w	Wind

Superscripts

$(\dot{\ })$	Per unit time
$(\ )''$	Per unit area
$(\dot{\ })''$	Per unit area, per unit time
$*$	Nondimensional

CHAPTER 1.
INTRODUCTION

1.1   Purpose

This report supplements and updates NUREG 1805, *Fire Dynamics Tools (FDT^S), Quantitative Fire Hazard Analysis Methods for the U.S. Nuclear Regulatory Commission Fire Protection Inspection Program.* The dynamic nature of fire is a quantitative and mathematically complex subject. It combines physics, chemistry, mathematics, and engineering principles and can be difficult to comprehend for those who have a limited background in these areas. The Fire Dynamics Tools (FDT^S) were developed to assist fire protection inspectors and others in solving fire hazard problems in nuclear power plants (NPPs). NUREG-1805 and the related Fire Dynamics Tools (FDT^S) provide first-order quantitative methods (i.e., traditional approaches, correlations, computations, closed form approximations or exact solutions, and hazard models) to assess the potential fire hazard development in commercial NPPs.

1.2   Organization of this Report

Since the publication of NUREG-1805 in December 2004, numerous comments and suggestions have been received from users of the spreadsheets. In response to these comments, all of the existing NUREG-1805 spreadsheets were updated to improve layout and printing. In addition, some spreadsheets have been revised to correct calculation errors and a new model has been added to the FDT^S suite. Table 1 lists the Fire Dynamics Tools and the associated chapter and related calculation methods. Table 2 identifies the spreadsheets that have been modified to correct calculation errors in addition to layout improvements.

Table 1-1. List of Fire Dynamics Tools

FDT^S* *indicated revised spreadsheet	Chapter and Related Calculation Method(s)
02.1_Temperature_NV_Sup1.xls 02.2_Temperature_FV_Sup1.xls* 02.3_Temperature_CC._Sup1.xls*	**Chapter 2.** Predicting Hot Gas Layer Temperature and Smoke Layer Height in a Room Fire with Natural and Forced Ventilation Method of McCaffrey, Quintiere, and Harkleroad (MQH) • Natural Ventilation Method of Foote, Pagni, and Alvares (FPA) • Forced Ventilation Method of Deal and Beyler • Forced Ventilation Method of Beyler • Fire in a Compartment with a Door Closed but with Sufficient Leaks to Prevent Pressure Buildup; Leakage is Ignored

FDT^{s}* *indicated revised spreadsheet	Chapter and Related Calculation Method(s)
03_HRR_Flame_Height_Burning_Duration_Calculation_Sup1.xls*	**Chapter 3.** Estimating Burning Characteristics of Liquid Pool Fire, Heat Release Rate, Burning Duration and Flame Height
04_Flame_Height_Calculations_Sup1.xls	**Chapter 4.** Estimating Wall Fire Flame Height, Line Fire Flame Height Against the Wall, and Corner Fire Flame Height
05.1_Heat_Flux_Calculations_Wind_Free_Sup1.xls	**Chapter 5.** Estimating Radiant Heat Flux from Fire to a Target Fuel *Wind-Free Condition* • Point Source Radiation Model (Target at Ground Level) • Solid Flame Radiation Model (Target at Ground Level) • Solid Flame Radiation Model (Target Above Ground Level) *Presence of Wind* • Solid Flame Radiation Model (Target at Ground Level) • Solid Flame Radiation Model (Target Above Ground Level)
05.2_Heat_Flux_Calculations_Wind_Sup1.xls*	
05.3_Thermal_Radiation_From_Hydrocarbon_Fireballs_Sup1.xls	Estimating Thermal Radiation from Hydrocarbon Fireballs
06_Ignition_Time_Calculations_Sup1.xls	**Chapter 6.** Estimating the Ignition Time of a Target Fuel Exposed to a Constant Radiative Heat Flux • Method of Estimating Piloted Ignition Time of Solid Materials Under Radiant Exposures Method of (1) Mikkola and Wichman, (2) Quintiere and Harkleroad, and (3) Janssens • Method of Estimating Piloted Ignition Time of Solid Materials Under Radiant Exposures Method of Toal, Silcock and Shields • Method of Estimating Piloted Ignition Time of Solid Materials Under Radiant Exposures Method of Tewarson
07_Cable_HRR_Calculations_Sup1.xls	**Chapter 7.** Estimating Full-Scale Heat Release Rate of a Cable Tray Fire
08_Burning_Duration_Soild_Sup1.xls	**Chapter 8.** Estimating Burning Duration of Solid Combustibles

FDT^s* *indicated revised spreadsheet	Chapter and Related Calculation Method(s)
09_Plume_Temperature_ Calculations_Sup1.xls	**Chapter 9.** Estimating Centerline Temperature of a Buoyant Fire Plume
10_Detector_Activation_Time_Sup1.xls*	Estimating Detector Response Time **Chapter 10.** Estimating Sprinkler Response Time **Chapter 11.** Estimating Smoke Detector Response Time **Chapter 12.** Estimating Heat Detector Response Time
13_Compartment_ Flashover_ Calculations_Sup1.xls	**Chapter 13.** Predicting Compartment Flashover • Compartment Post-Flashover Temperature: Method of Law • Minimum Heat Release Rate Required to Compartment Flashover: Method of (1) McCaffrey, Quintiere, and Harkleroad (MQH); (2) Babrauskas; and (3) Thomas
14_Compartment_Over_Pressure_ Calculations_Sup1.xls	**Chapter 14.** Estimating Pressure Rise Attributable to a Fire in a Closed Compartment
15_Explosion_Claculations_Sup1.xls	**Chapter 15.** Estimating the Pressure Increase and Explosive Energy Release Associated with Explosions
16_Battery_Room_Flammable_Gas_ Conc_Sup1.xls*	**Chapter 16.** Calculating the Rate of Hydrogen Gas Generation in Battery Rooms • Method of Estimating Hydrogen Gas Generation Rate in Battery Rooms • Method of Estimating Flammable Gas and Vapor Concentration Buildup in Enclosed Spaces • Method of Estimating Flammable Gas and Vapor Concentration Buildup Time in Enclosed Spaces

FDT^s*	Chapter and Related Calculation Method(s)
*indicated revised spreadsheet	
17.1_FR_Beams_Columns_ Substitution_Correlation_Sup1.xls*

17.2_FR_Beams_Columns_Quasi_ Steady_State_Spray_Insulated_Sup1.xls*

17.3_FR_Beams_Columns_Quasi_ Steady_State_Board_Insulated_Sup1.xls*

17.4_FR_Beams_Columns_Quasi_ Steady_State_Uninsulated_Sup1.xls* | **Chapter 17.** Calculating the Fire Resistance of Structural Steel Members

• Empirical Correlations

• Beam Substitution Correlation (Spray-Applied Materials)
• Column Substitution Correlation (Spray-Applied Materials)
• Heat Transfer Analysis using Numerical Methods Protected Steel Beams and Columns (Spray-Applied)

• Heat Transfer Analysis using Numerical Methods Protected Steel Beams and Columns (Board Materials)

• Heat Transfer Analysis using Numerical Methods Unprotected Steel Beams and Columns |
| 18_Visibility_Through_Smoke_Sup1.xls | **Chapter 18.** Estimating Visibility Through Smoke |
| 19_THIEF_of_Cables_Calculations_ Sup1.xls | **Chapter 19.** Estimating the Thermally-Induced Electrical Failure (THIEF) of Cables |

Table 1-2. Summary of Technical Changes to Original NUREG-1805 Spreadsheets

CHAPTER TITLE	
Spreadsheet Title	**Description of Revision**
CHAPTER 2. PREDICTING HOT GAS LAYER TEMPERATURE IN A ROOM FIRE WITH FORCED VENTILATION	
02.2_Temperature_FV_Sup.xls	Deleted extraneous variable in Cell C94. Under Deal & Beyler Method, added t_p calculation in Cells starting B148, added h_k formula for $t > t_p$ in Cell E158 and changed equation for results table for h_k in Cells starting D184.
02.3_Temperature_CC_Sup1.xls	Corrected calculation for Parameter K_1. Added area of compartment enclosing surface boundaries (A_T) in Parameter K_1.
CHAPTER 3. ESTIMATING BURNING CHARACTERISTICS OF LIQUID POOL FIRE, HEAT RELEASE RATE, BURNING DURATION, AND FLAME HEIGHT	
03_HRR_Flame_Height_Burning_Duration_Calculations_Sup1.xls	Corrected diameter reference in Q calculation column in table.
CHAPTER 5. ESTIMATING RADIANT HEAT FLUX FROM FIRE TO A TARGET FUEL AT GROUND LEVEL IN PRESENCE OF WIND (TILTED FLAME) SOLID FLAME RADIATION MODEL	
05.2_Heat_Flux_Calculations_Wind_Sup1.xls	Corrected typo in Cell C85. Corrected programming errors in Cells G156 and M158.
CHAPTER 10. ESTIMATING SPRINKLER RESPONSE TIME	
10_Detector_Activation_Time_Sup1.xls (Sprinkler)	Changed references from Qc to Q and 0.15 to 0.18 in cells B58, B59, B76, C62, C78, D74, D76, E58, and E59, as appropriate and changed calculations in cells C76, C79 and E76.
CHAPTER 11. ESTIMATING SMOKE DETECTOR RESPONSE TIME	
10_Detector_Activation_Time_Sup1.xls (Smoke_Detector)	Changed references from Q_c to Q and 0.15 to 0.18 in cells B37, B38, B55, C41, C57, D53, D55, E37, and E38, as appropriate and changed calculations in cells C55, C58 and E55.
CHAPTER 12. ESTIMATING HEAT DETECTOR RESPONSE TIME	
10_Detector_Activation_Time_Sup1.xls (FTHDetector)	Changed references from Qc to Q and 0.15 to 0.18 in cells B98, B99, B116, C102, C118, D114, D116, E98, and E99, as appropriate and changed calculations in cells C116, C119 and E116.
CHAPTER 16. CALCULATING THE RATE OF HYDROGEN GAS GENERATION IN BATTERY ROOMS	
16_Battery_Room_Flammable_Gas_Conc_Sup1.xls	Corrected Filed D61 air flow rate (fresh air) in enclosure calculation.

CHAPTER TITLE	
Spreadsheet Title	**Description of Revision**
CHAPTER 17. ESTIMATING FIRE RESISTANCE TIME OF STEEL BEAMS PROTECTED BY FIRE PROTECTION INSULATION (QUASI-STEADY-STATE APPROACH)	
17.3_FR_Beams_Columns_Quasi_ Steady_State_Board_Insulated_Sup1.xls	Change reference from I16 to I15, add variable ambient temperature, and corrected ambient temperature reference in Results calculations.
17.3_FR_Beams_Columns_Quasi_ Steady_State_Board_Insulated_Sup1.xls	Change reference from I16 to I15, add variable ambient temperature, and corrected ambient temperature reference in Results calculations.
17.4_FR_Beams_Columns_Quasi_ Steady_State_Board_Insulated_Sup1.xls	Corrected Typo in Equation $T_f = C_1 \, LOG \, (0.133 \, t + 1) + T_a$ to Calculate Fire Exposure Temperature. Corrected Conversion from Kelvin to Fahrenheit in Result Field.
17.4_FR_Beams_Columns_Quasi_ Steady_State_Board_Insulated_Sup1.xls	Corrected Typo in Equation $T_f = C_1 \, LOG \, (0.133 \, t + 1) + T_a$ to Calculate Fire Exposure Temperature. Corrected Conversion from Kelvin to Fahrenheit in Result Field.

Chapters 1 through 18 from NUREG-1805 are not duplicated in this report. However, example problems that have been solved using the revised NUREG-1805 spreadsheets are included in Appendix A (English Units) and Appendix B (S.I. Units) of this report. The reader is referred to the original NUREG-1805 report for technical documentation of those spreadsheets. The chapter describing the new model is labeled as Chapter 19 continuing in the FDT^s series of chapters. The CD-ROM accompanying this report contains the new and updated spreadsheets. Spreadsheets which use English units for the majority of the input data are located in the folder labeled "English Units." The folder labeled "SI Units" contains the same FDT^s spreadsheets but requires user inputs to be in SI units.

CHAPTER 19.
ESTIMATING THE THERMALLY-INDUCED ELECTRICAL FAILURE (THIEF) OF CABLES

19.1   Objectives

This chapter has the following objectives:

- Describe the three primary functions performed by electrical cables in nuclear power plants (NPPs): control, instrumentation, and power.

- Identify the two major types of insulation used for cables: thermoset (TS) and thermoplastic (TP).

- Explain the factors that determine how a cable will respond to fire exposure.

- Explain the processes that electrical failures can initiate in a cable tray.

19.2   Introduction

Electrical cables perform numerous functions in NPPs. For example, power cables supply electricity to motors, transformers, heaters, and light fixtures. Control cables connect plant equipment such as motor-operated valves (MOVs) and motor starters to remote initiating devices (e.g., switches, relays, and contacts). Instrumentation cables transmit low-voltage signals between input devices and readout display panels.

NPPs typically contain hundreds of miles of electrical cables. A typical boiling-water reactor (BWR) requires about 97 km (60 miles) of power cable, 80.5 km (50 miles) of control cable, and 402 km (250 miles) of instrument cable. A pressurized-water reactor (PWR) may require even more cables. The containment building of Waterford Steam Electric Generating Station, Unit 3 requires nearly 1,609 km (1,000 miles) of cable (NUREG/CR-6384).

Electrical cables have been responsible for a number of fires in NPPs over the years. A 1966 National Fire Protection Association study (Hedland, 1966) identified 24 fires with the most serious at that time occurring at the Peach Bottom Atomic Power Station operated by Philadelphia Electric Company. In 1975, a fire involving electrical cables occurred at the Browns Ferry Nuclear Power Plant operated by the Tennessee Valley Authority (NUREG-0050). The fire caused damage to more than 1,600 cables resulting in significant loss of emergency core cooling system equipment. In addition to loss of power to motors, valves, and other equipment, fire-induced short circuits caused many instrument, alarm, and indicating circuits to produce false and conflicting indications of equipment operation.

The behavior of cables in a fire depends on a number of factors including their constituent materials and construction as well as their location and installation geometry. Burning cables can propagate flames from one area to another or they can add to the amount of fuel available for combustion and can liberate smoke containing toxic and corrosive gases. Fire exposure of an electrical cable can cause a loss of insulation resistance, loss of insulation physical integrity (i.e., melting of the insulation), and electrical breakdown or short-circuiting. The lower the heat flux required to ignite the electrical cables, the greater the fire hazard in terms of ignition and flame spread.

Fire-induced damage to a cable can result in one of the following failure modes (LaChance et al., 2000):

- Open Circuit: loss of electrical continuity of an individual conductor (i.e., the conductor is broken and the signal or power does not reach its destination).

- Short to Ground: an individual conductor comes into electrical contact with a grounded conducting medium (such as a cable tray, conduit, or a grounded conductor) resulting in a low-resistance path that diverts current from a circuit. The fault may be accompanied by a surge of excess current to ground (particularly in higher voltage circuits) that is often damaging to the conductor.

- Hot Short: electrical faults that involve an energized conductor contacting another conductor of either the same cable (a conductor-to-conductor hot short) or an adjacent cable (a cable-to-cable hot short). A hot short has the potential to energize the affected conductor or to complete an undesirable circuit path.

In addition to these failure modes, the wide variations in the composition of polymer materials used in electrical cable insulation has resulted in notable failure characteristics among the two broad classes of polymer materials—thermoset (TS) and thermoplastic (TP). Thermoplastic cables are characterized by softening, melting, and dripping when exposed to a damaging thermal environment. Thermoset insulated cables don't have a distinct melting point. As a result of their molecular bonding, TS cable degradation is characterized by voiding, off gassing, and swelling.

TS and TP cables not only fail differently, but their use in different types of circuits can cause unique circuit response. For instance, instrumentation circuits typically operate using a 4-20 mA signal, where 4 mA equals 0 percent of the parameter being measured (e.g., pressure, flow, volume, etc.) and 20 mA equals 100 percent. Testing has shown that a TP instrument cable will fail abruptly while a TS cable will fail gradually. This behavior in an instrument circuit could provide an operator with misleading information (e.g., loss of pressure or level). The operator would not be able to readily determine if the readout was an accurate indication of plant status or if it was the result of fire damage to a cable. A prolonged transition from "good signal" to obviously "faulty signal" proved to be typical of the TS cables. Figure 19-1 illustrates the different response characteristics of TS and TP cables when exposed to a fire environment.

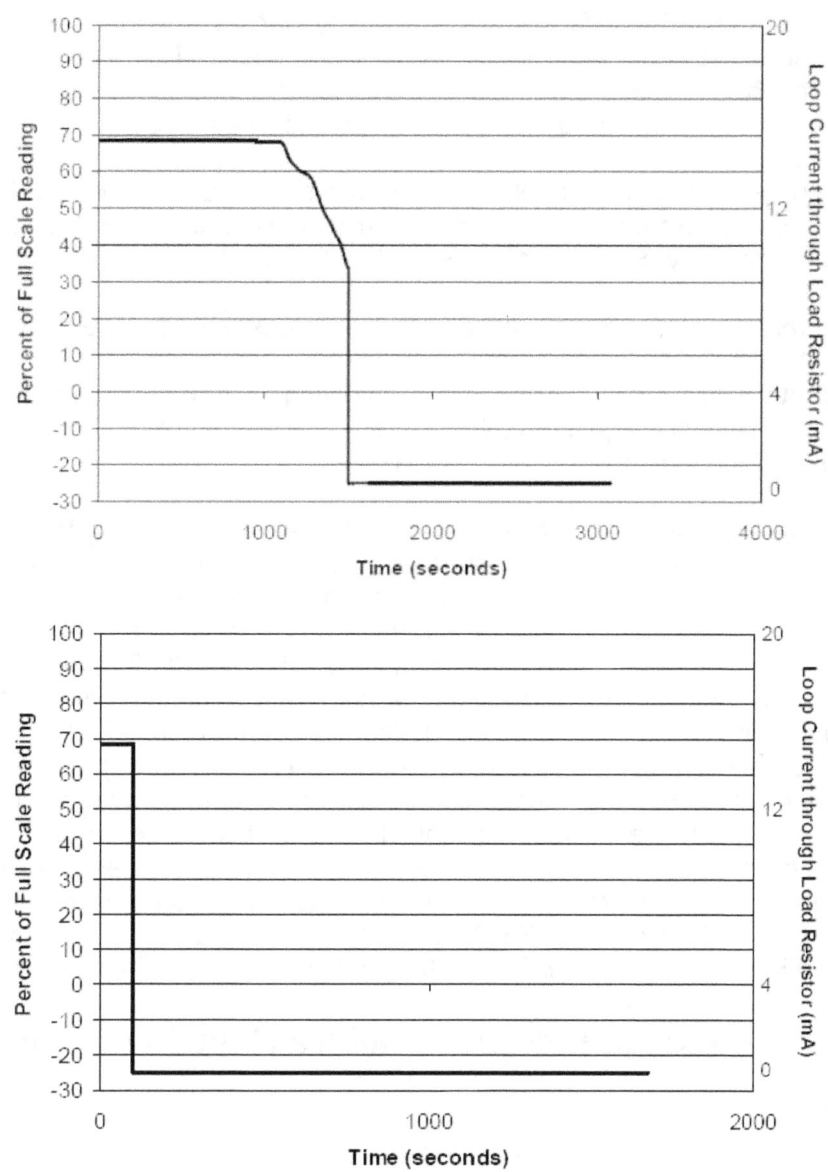

Figure 19-1.   Different signal responses from a thermoset cable (top) and a thermoplastic cable (bottom) when exposed to a fire.

A key NPP application is the Fire Probabilistic Risk Assessment (FPRA), which often relies on fire models to predict cable failure times for a predefined set of fire conditions (NUREG/CR 6850, Vol. 1). To assess the conditional probability of cable damage given a fire, these failure times are weighed against the likelihood that fire suppression succeeds within the available time. The ability of current compartment fire models to predict cable damage is limited. For example, in the NIST Consolidated Model of Fire and Smoke Transport (CFAST) (Jones, et al., 2005), a general thermal target response submodel is available, but this model was not specifically developed for, nor has it been calibrated for, cables as the thermal target. Hence, one primary need with respect to fire model improvement is the development, calibration, and validation of predictive thermal/damage target response models specific to cables as the target.

In late 2006, a series of cable fire tests were performed by Sandia National Laboratories (SNL) under the sponsorship of the U.S. Nuclear Regulatory Commission (NRC) Office of Nuclear Regulatory Research (RES). This program, known as the CAble Response tO Live FIRE (CAROLFIRE) project, was designed to address two specific need areas:

1. Provide an experimental basis for resolving five of the six items identified as "Bin 2" circuit configurations in Risk-informed Approach for Post-Fire Safe-Shutdown Circuit Inspections, Regulatory Issue Summary (RIS) 2004-03, Rev. 1, 12/29/04.

2. Improve fire modeling tools for the prediction of cable damage under fire conditions.

The project was conducted as a collaborative effort involving representatives of RES, SNL, the NRC Office of Nuclear Reactor Regulation (NRR), the National Institute of Standards and Technology (NIST), and the University of Maryland (UMd).

The CAROLFIRE project included a series of 78 small-scale radiant heating tests and 18 intermediate-scale open burn tests. The small-scale tests were performed in an SNL facility called Penlight and involved exposure of two to seven lengths of cable to grey-body radiant heating. These tests were aimed in large part at the fire model improvement need area, but they also provided data pertinent to the resolution of two of the five Bin 2 items. The intermediate-scale tests involved exposure of cables, generally in bundles of 6 to 12 cables, under various routing configurations and at various locations within a relatively open test structure. The fires were initiated by a propene (also known as propylene) gas diffusion burner. The fire typically ignited, at a minimum, those cables located directly above the fire source. The tested cables were representative of those currently in use at U.S. commercial NPPs. Testing included a broad range of both TS and TP insulated cables as well as one mixed TS-insulated and TP-jacketed cable. The three-volume test report includes two volumes documenting the test results (NUREG/CR-6931 Vol. 1, NUREG/CR-6931 Vol. 2) and a volume describing the development of the THIEF model (NUREG/CR-6931 Vol. 3).

19.3   Thermally-Induced Electrical Failure (THIEF) of Cables

19.3.1   General

The development of a predictive model of cable failure has been elusive for a number of reasons. First, cables are a fairly complex combination of insulating plastics, metal conductors, protective armors, and a variety of filler materials. The availability of comprehensive thermo-physical properties of these materials is limited. Even when the material properties for a particular cable are available, it is still a challenge to calculate the heat penetration through a bundle of the cables lying in a tray or run through a conduit. Rather than try to develop detailed models, a simpler approach is to develop an empirical relationship between the time to electrical failure and the "exposing" temperature (i.e., the temperature of the hot gases in the vicinity of the cable).

In Appendix A of NUREG-1805, a set of engineering calculation methods specifically designed for nuclear power plant applications suggests that the time to electrical failure is inversely proportional to the exposing temperature. For the two major classes of cables, TS and TP, it provides an estimated failure time for a given exposing temperature. Although these equations are useful screening methods, they have some significant limitations. First, the equations are based on constant temperature exposures, which is unrealistic in a fire scenario. Second, they do not account for different cable installations or configurations. The formulae only distinguish between a TS and TP cable based on the latter having been shown to fail at lower temperatures than the former. The formulae do not take into account size, mass, protective barriers, or site-specific conditions. A more flexible predictive model should have some consideration for the thermal mass of the cable, and it must infer electrical failure from the attainment of a given "failure" temperature somewhere within the cable.

Petra Andersson and Patrick Van Hees at the Swedish National Testing and Research Institute (SP) proposed that a cable's thermally-induced electrical failure (THIEF) can be predicted via a simple one-dimensional heat-transfer calculation under the assumption that the cable can be treated as a homogenous cylinder (Andersson and Van Hees, 2005). Their results for PVC cables were encouraging and suggested that the simplified analysis could be used for other types of cables. Using the data from the CAROLFIRE project, researchers at NIST examined the feasibility of extending the THIEF model to other cable types (NUREG/CR-6931 Vol. 3).

The governing equation for the cable temperature $T(r,t)$ is given by:

$$\rho c \frac{\partial T}{\partial t} = \frac{1}{r}\frac{\partial}{\partial r}k\,r\,\frac{\partial T}{\partial r} \tag{19-1}$$

where ρ, c, and k are the effective density, specific heat, and thermal conductivity of the solid, respectively. The boundary condition at the exterior boundary, $r = R$ is given by:

$$k\frac{\partial T}{\partial r}(R,t) = \dot{q}'' \tag{19-2}$$

where $\dot{q}''$ is the assumed axially symmetric heat flux to the exterior of the cable. The heat flux is provided by the fire model or fire analysis that is being used to assess the overall thermal environment of the compartment where the cable is located. In most realistic fire scenarios, the heat flux to the cable is not axially symmetric. For the purpose of modeling the cable failure, it is recommended that the maximum value be used.

While Andersson and Van Hees developed an analytical solution for their THIEF model, the NIST researchers proposed a numerical solution as simpler and easier to implement in fire models. To solve Eq. (19-1) numerically, the radius, R, of the cable is divided into N uniformly spaced increments of length, $\delta r = R/N$. An appropriate value for δr is about 0.1 mm for cables similar to those tested in CAROLFIRE. Next, a time step is defined that is related to the spatial increment. This is known as the *time step constraint*, which is necessary for accuracy and sometimes numerical stability:

$$\delta t = \frac{c\,\rho\,\delta r^2}{2k} \tag{19-3}$$

The temperature of the i-th radial increment (or *cell*) at the n-th time step ($t^n = n\delta t$) is denoted, T_i^n. The value of the radius at the forward edge of the i-th cell is denoted, $r_i = i\delta r$. Thus, $r_0 = 0$ and $r_N = R$.

A finite difference approximation to equation 19-1 is given by:

$$T_i^{n+1} = T_i^n + \frac{\delta t\, \alpha}{r_{avg}\, \delta r}\left[r_i \frac{\left(T_{i+1}^n - T_i^n\right)}{\delta r} - r_{i-1} \frac{\left(T_i^n - T_{i-1}^n\right)}{\delta r} \right]$$ (19-4)

where α is the thermal diffusivity of the cable material (m^2/s).

The THIEF model predicts the temperature profile within the cable as a function of time, given a time-dependent exposing temperature or heat flux. The CAROLFIRE experimental program included bench-scale, single-cable experiments in which temperature measurements were made on the surface of, and at various points within, cables subjected to a uniform heat flux. These experiments provide the link between internal cable temperature and electrical failure. The model infers electrical failure when a given "failure" temperature is reached based on the calculated interior temperature. The temperature of the centermost point in the cable is not necessarily the indicator of electrical failure. This analysis method uses the temperature just inside the cable jacket rather than the centermost temperature because that is where electrical shorts in a multi-conductor cable are most likely to occur first.

The THIEF model uses the general cable construction and bulk properties but does not require more detailed thermo-physical properties. For example, the mass per unit length and diameter are needed, but the thermal conductivity, specific heat, and emissivity are assumed based on the current generation of cables in existing plants. This latter detailed information is not always readily available for the wide variety of often proprietary cable materials, and bench-scale experiments to measure the properties can be expensive and difficult to perform for all existing and future cable materials.

19.3.2   Cable Properties and Selection Process

To determine the time of cable failure during thermal insult, the THIEF model requires three cable properties:

- Cable outside diameter (OD).
- Mass per unit length.
- Jacket thickness.

The THIEF spreadsheet allows the user to enter these parameters in two different ways. If the three cable parameters are known, the user can enter them directly into the THIEF spreadsheet. When the cable parameters are not known, the user can use the automated cable selection process built into the spreadsheet.

To support the identification of cable properties, an extensive data-gathering effort was conducted to identify these properties for the most common cable types used in an NPP (i.e., power, control, and instrumentation cables). Over 3,000 cables from nine different major cable manufactures were identified. Cables are manufactured in various sizes, conductor counts, and material types to provide the appropriate circuit function for the particular application. These variations result in thousands of different cable designs being manufactured to support specific needs. As a result, it is nearly impossible to capture every single cable type that exists.

The cable parameters gathered for this project are largely based on cable constructions typically found in an NPP, and the information is as accurate as what is presented on the manufacturer's website. The following manufacturer's cable specifications are included in the spreadsheet:

- American Insulated Wire.
- Cable USA.
- Continental.
- Dekoron.
- Draka.
- First Capital.
- General Cable.
- Okonite.
- Rockbestos-Surprenant.

The THIEF spreadsheet has been automated to sort the complete list of over 3,000 cables by characteristics that the user specifies. To perform this sorting function, the user first clicks on the "Select Cable" button located in the input parameters section of the spreadsheet. When a new window opens, click on the "Select Cable From List" choice. The next screen allows the user to select the general characteristics of the cable (i.e., cable type [control, instrument, or power], wire gauge [AWG], and the number of conductors). The THIEF model uses these three input parameters to sort the complete cable list and narrow it down to the cables that meet the user's input parameters. The condensed list is displayed and sorted by manufacturer, cable model, insulation, and jacket material. The user then selects the cable to be analyzed from the list.

The list is not intended to be inclusive of all cables, and there will be instances were a specific cable of concern is not listed. In cases where the cable of interest is not listed, the user may (1) look up the information from the cable manufacturer's Web site, (2) get the information from the license documentation and enter the values directly into the spreadsheet, or (3) choose a "generic" cable from the list with similar construction and insulation materials. Based on an analysis of manufacturer data, "generic" cables have been identified for each insulation type, cable function, number of conductors, and wire gauge.

When choosing a cable from the list, it is important to choose cable parameters that are representative of the cables found in the plant. For example, if a cable located in the plant has the identification, "GENERAL CABLE® EHTC 7/C 12 AWG EPR/HYP UL TYPE TC 90°C," it is important to choose a cable that has the same general characteristics. That is, choose a cable with seven conductors (7/C) with each conductor size 12 AWG and with the same insulation and jacket materials. Choosing the exact manufacture is less important than physical properties; however, if the exact cable is found in the list, this is the most appropriate choice. Some types of instrumentation cables may be identified in terms of numbers of pairs or numbers of triads. In these cases, the actual number of conductors should be used as input for the spreadsheet. For example, a two-pair cable would be analyzed as a four-conductor cable. A three-triad cable would have nine conductors. In cases where the specific cable to be analyzed is not listed in the condensed cable list, the user may select a cable with identical cable materials (PVC, XLPE, SR, etc.) or an insulation material of the same type (TP or TS). Table 19-1 presents a grouping of various cable insulation materials into the two classes.

Table 19-1. Common Thermoplastic and Thermoset Cable Insulation and Jacket Materials

Thermoplastic (TP)	Thermoset (TS)
ETFE (Tefzel)	SR (Silicon Rubber)
FEP (Teflon)	EPR (Ethylene Propylene Rubber)
CPE (Chlorinated Polyethylene)	EPDM (Ethylene Propylene diene monomer rubber)
PE (Polyethylene)	CP (Neoprene)
PVC (Polyvinyl Chloride)	CSPE (Chlorosulphonated polyethylene)
	XLPE (Cross-linked Polyethylene)
	XLPO (Cross-linked Polyolofin)

The last input parameter needed from the user is information related to where the cable is located (air drop, cable tray, or conduit). When a cable is suspended in air (air drop) or in a cable tray, a majority of the cable will be exposed to the fire environment. A cable in conduit or armored cable will be "shielded" to some extent from direct exposure to the fire environment. Once the cable location is selected, the cable parameters are automatically entered into the THIEF spreadsheet, including the failure temperature, based on the cables insulation type (TP or TS).

19.3.3 Exposure Gas Temperature

The exposure gas temperature can be obtained in several ways. First, the user can manually enter a set of data points obtained from the results of a fire model such as CFAST, MAGIC, or FDS. Alternatively, the data points could be obtained from measurements recorded during a fire test representative of the scenario. Finally, the THIEF model spreadsheet has been integrated with several spreadsheets previously issued in NUREG 1805. These spreadsheets are appropriate for determining hot gas layer temperature (Chapter 2.6) or plume centerline temperature (Chapter 9). Specifically, the following methods are included:

- McCaffrey, Quintiere, and Harkleroad (MQH) for natural ventilation (2.6.1).
- Beyler for natural ventilation in a closed compartment (2.6.2).
- Foote, Pagni, and Alvares (FPA) for forced ventilation (2.6.3).
- Deal and Beyler for forced ventilation (2.6.4).
- Centerline temperature of a buoyant plume (9).

These spreadsheets can be used to calculate the exposure gas temperature profile, and the results will be automatically incorporated into the THIEF model calculation.

Typically, a cable will be in one of four locations relative to the fire at any given time. Figure 19-2 illustrates the four possible orientations for the cable and fire. The cable can be (a) in the flames, (b) in the smoke plume, (c) in the hot gas layer, or (d) in the "ambient" lower layer. The dashed lines in Figure 19-2 represent the radiant energy from the fire. Typically, a small amount of the radiant flux reaching the hot gas layer will be absorbed (smaller arrows in Figure 19-2) while the remainder is "reflected" back to the items in the lower layer. The object labeled "cable tray" in Figure 19-2 could represent a cable tray, an air drop, or a conduit. The exposure gas temperature will depend on the assumed location of the cables relative to the fire at the time of interest. As the fire develops, the location of the cables relative to the fire may change. So, multiple calculations may be necessary to evaluate the potential for cable failure.

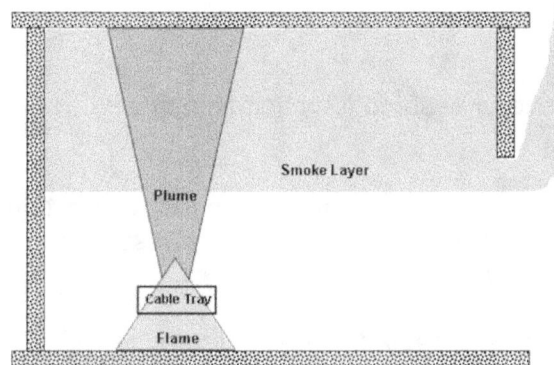

(a) Cable within flames from fire.

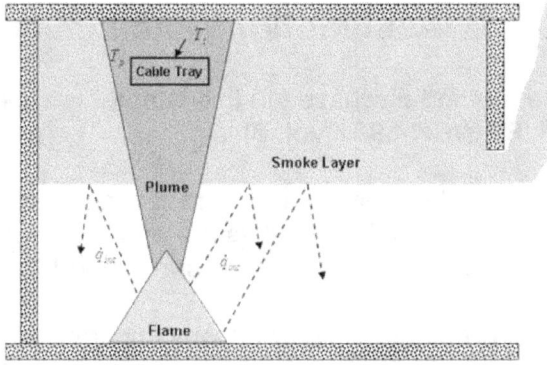

(b) Cable within smoke plume from fire.

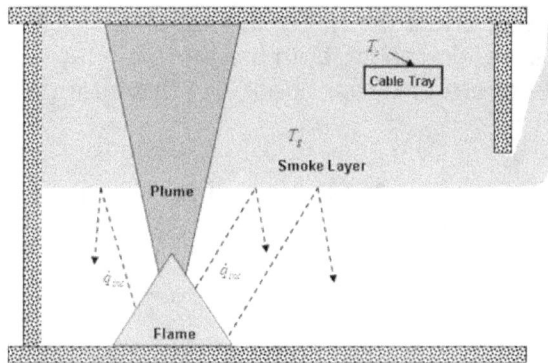

(c) Cable in hot gas layer.

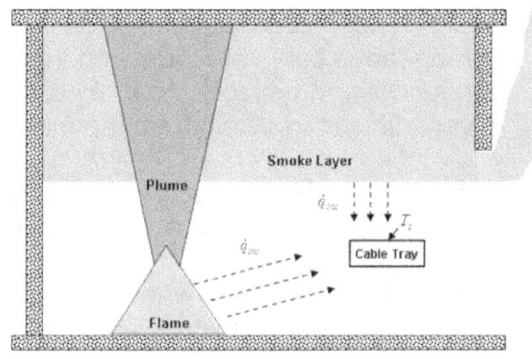

(d) Cable in "ambient" lower layer.

Figure 19-2.   Drawing showing the four possible locations for a cable tray relative to the fire.

When the cable is within the flames from the fire (Figure 19-2[a]), the cable is assumed to fail upon flame contact. If it needs to function, it does not; if it could have "Hot Shorted" and made bad things happen, it did. The cable is on fire or shortly will be on fire. When the cable is within the smoke plume directly above the fire (Figure 19-2[b]), the temperature exposure is not as severe as when it is within the flames. The temperature at any location within the fire plume can be calculated using the spreadsheet described in Chapter 9 of NUREG-1805. When the cable is in the hot gas layer (Figure 19-2[c]), the fire is located in another part of the room not directly under the cable. The fire will "pump" hot gas into the upper part of the room. This gas will be contained in the room until the interface (the separation between the hot gas layer and the "ambient" lower layer) drops below the top level of some opening, such as an open door or window. When the cable is located with the hot gas layer, the exposure gas temperature will be equal to the hot gas layer temperature. Chapter 2 of NUREG 1805 describes several spreadsheets that calculate hot gas layer temperature depending on room openings and ventilation conditions. In the final scenario (Figure 19-2[d]), the hot gas layer never reaches the cables of interest. The cables remain in the lower "ambient" temperature layer. In this case, the cable will be heated solely by the radiant energy from the fire. The THIEF model is not directly applicable to this scenario because there is no surrounding hot gas layer. However, the cable could still fail from the radiant energy exposure. NUREG 1805 includes spreadsheets in Chapter 5 for calculating the radiant exposure to the cables and guidance on selecting radiant exposure levels that could lead to cable failure.

19.4   Assumptions and Limitations

The method discussed in this chapter is subject to several assumptions and limitations (NUREG/CR-6931 Vol. 3):

(1)   The heat penetration into a cable of circular cross section is largely in the radial direction. This greatly simplifies the analysis, and it is also conservative because it is assumed that the cable is completely surrounded by the heat source.

(2)   The cable is homogenous in composition. In reality, a cable is constructed of several different types of polymeric materials, cellulose fillers, and a conducting metal, most often copper.

(3)   The thermal properties—conductivity, specific heat, and density—of the assumed homogenous cable are independent of temperature. In reality, both the thermal conductivity and specific heat of polymers are temperature-dependent, but this information is very difficult to obtain from the manufacturers.

(4)   It is assumed that no decomposition reactions occur within the cable during its heating, and ignition and burning are not considered in the model.

(5)   Electrical failure occurs when the temperature just inside the cable jacket reaches an experimentally determined value.

(6)   The temperature in the upper layer is uniform throughout the layer regardless of location within the layer or distance from the fire.

19.5   Required Input for Spreadsheet Calculations

The user must obtain the following information before using the spreadsheet:

(1)   Gas temperature around the cable as a function time (from fire models, fire tests, or integrated spreadsheets).

(2)   Type of cable (thermoplastic, thermoset).*

(3)   Cable diameter.*

(4)   Cable mass per unit length.*

(5)   Cable jacket thickness.*

(6)   Conduit thickness, if any.

(7)   Conduit outer diameter, if any.

* Can be obtained from list of cable properties integrated into THIEF spreadsheet when cable function (power, control, or instrumentation), wire size (AWG), and number of conductors are known.

19.6   Cautions

(1)   Use the (19_THIEF_Thermally_Induced_Electrical_Failure_of_Cables.xls) spreadsheet on the CD-ROM.

(2)   Make sure to enter the input parameters in the correct units.

(3)   The failure temperature is an estimate based on data from a number of different cable tests. Although a single value is used to estimate failure, the graph shows the temperature range over which failure may occur for the selected cable type (TP or TS).

19.7   Summary

Determining the failure time of cables exposed to a hot gas layer involves the following steps:

(1)   Determine the exposure gas temperature using test data, fire model results, or incorporated spreadsheets.

(2)   Select cable of interest from one of the lists or enter user determined cable properties.

(3)   Calculate temperature profile through cable as a function of time and hot gas exposure.

(4)   Determine when failure temperature is reached.

19.8   Decision Criteria

The failure temperature range is estimated from temperature measurements obtained just below the cable jacket during the CAROLFIRE tests. For TP cables, the temperatures reached somewhere between 200 °C (390 °F) and 250 °C (480 °F). For TS cables, the range was about 400 °C (750 °F) to 450 °C (840 °F).

19.9   References

Andersson, P. and P. Van Hees, "Performance of Cables Subjected to Elevated Temperatures," *Fire Safety Science – Proceedings of the Eighth International Symposium*, International Association for Fire Safety Science, 2005.

Hedlund, C.F., "Grouped Combustible Wire and Cables," *Fire Journal*, Volume 60, pp. 5–8, March 1966.

Jones, W. W., R.D. Peacock, G.P. Forney, and P.A. Reneke, "Consolidated Model of Fire Growth and Smoke Transport (Version 6): Technical Reference Guide," NIST SP 1026, National Institute of Standards and Technology, Gaithersburg, MD, 2005.

LaChance, J., S.P. Nowlen, F. Wyant, and V. Dandini, "Circuit Analysis-Failure Mode and Likelihood Analysis," A Letter Report to USNRC, Sandia National Laboratory, Albuquerque, New Mexico, May 8, 2000.

NUREG-0050, "Recommendations Related to Browns Ferry Fire," U.S. Nuclear Regulatory Commission, Washington, DC, February 1976.

NUREG-1805, *Fire Dynamics Tools (FDT^s) – Quantitative Fire Hazard Analysis Methods for the U.S. Nuclear Regulatory Commission Fire Protection Inspection Program*, U.S. Nuclear Regulatory Commission, Washington, D.C., 2004.

NUREG/CR-6384 Vol. 1, "Literature Review of Environmental Qualification of Safety-Related Electric Cables," U.S. Nuclear Regulatory Commission, Washington, DC, April 1996.

NUREG/CR-6850 Vol. 1, "EPRI/NRC-RES Fire PRA Methodology for Nuclear Power Facilities Volume 1: Summary & Overview," U.S. Nuclear Regulatory Commission, Washington, DC, September 2005.

NUREG/CR-6931 Vol. 1, "Cable Response to Live Fire (CAROLFIRE) Volume 1: General Test Descriptions and the Analysis of Circuit Response Data," U.S. Nuclear Regulatory Commission, Washington, DC, 2007.

NUREG/CR-6931 Vol. 2, "Cable Response to Live Fire (CAROLFIRE) Volume 2: Cable Fire Response Data for Fire Model Improvement," U.S. Nuclear Regulatory Commission, Washington, DC, 2007.

NUREG/CR-6931 Vol. 3, "Cable Response to Live Fire (CAROLFIRE) Volume 3: Thermally-Induced Electrical Failure (THIEF) Model," U.S. Nuclear Regulatory Commission, Washington, DC, 2008.

Risk-informed Approach for Post-Fire Safe-Shutdown Circuit Inspections," Regulatory Issue Summary (RIS) 2004-03, Rev. 1, U.S. Nuclear Regulatory Commission, Washington, DC, December 2004.

19.10   Additional Readings

Chavez, J.M., and Klamerus, L.J., "Cable Tray Fire Experimentation," LA-9911-C-Vol. 1; LA-9911-C-Vol. 2; CSNI No. 83; CSNI Specialist Meeting on Interaction of Fire and Explosion With Ventilation Systems in Nuclear Facilities, April 25-28, 1983, Los Alamos, New Mexico, pp. 375-390, 1983.

Dey, M.K., "Evaluation of Fire Models for Nuclear Power Plant Applications: Cable Tray Fires," NISTIR 6872, National Institute of Standards and Technology, Gaithersburg, Maryland, June 2002.

Klamerus, L.J., "Cable Tray Fire Tests," IEEE Power Generation Committee of the IEEE Power Engineering Society, IEEE Winter Meeting, February 4-9, 1979, New York, New York, pp. 1-12, 1979.

Lee, B.T., "Heat Release Rate Characteristics of Some Combustible Fuel Sources in Nuclear Power Plants," NBSIR 85-3195, National Bureau of Standards, Gaithersburg, Maryland, July 1985.

Nicolette, V.F., and Yang, K.T., "Modeling of HDR Oil and Cable Fire Tests," Structural Mechanics in Reactor Technology, 12th International Conference, Post Conference Seminar #6: Fire Safety of Nuclear Power Plants, August 23-24, 1993, Heidelberg, Germany, pp. 1-18, 1993.

NUREG/BR-0361, "The Browns Ferry Nuclear Plant Fire of 1975 and the History of NRC Fire Regulations," U.S. Nuclear Regulatory Commission, Washington, DC, February 2009.

NUREG/CR-2431, "Burn Mode Analysis of Horizontal Cable Tray Fires," U.S. Nuclear Regulatory Commission, Washington, DC, February 1982.

NUREG/CR-2927, "Nuclear Power Plant Electrical Cable Damageability Experiments," U.S. Nuclear Regulatory Commission, Washington, DC, October 1982.

NUREG/CR-4527 Vol. 2, "Experimental Investigation of Internally Ignited Fires in Nuclear Power Plant Control Cabinets, Part 2, Room Effects Tests," U.S. Nuclear Regulatory Commission, Washington, DC, November 1988.

NUREG/CR-4679, "Quantitative Data on the Fire Behavior of Combustible Materials Found in Nuclear Power Plants: A Literature Review," U.S. Nuclear Regulatory Commission, Washington, DC, February 1987.

NUREG/CR-6776, "Cable Insulation Resistance Measurements Made During Cable Fire Tests," U.S. Nuclear Regulatory Commission, Washington, DC, June 2002.

Pryor, A.J., "Browns Ferry Revisited," *Fire Journal*, Vol. 71, No. 3, pp. 85-86, 88-89, 120-121, May 1977.

Sawyer, R.G.; and Elsner, J.A., "Cable Fire at Browns Ferry Nuclear Power Plant," *Fire Journal*, Vol. 70, No. 4, pp. 5-10, July 1976.

Shen, T.S., "Will the Second Cable Tray Be Ignited In a Nuclear Power Plant?," *Journal of Fire Sciences*, Vol. 24, No. 4, 265-274, July 2006.

19.11 Problems

Example Problem 19.11-1 (SI Units)
(User Input - Cable Properties and Exposure Gas Temperature)

Problem Statement

As part of the development of the THIEF model, a number of experiments were conducted in the "Penlight" apparatus by Sandia National Laboratories. Figure 19-3 shows a schematic and photograph of the penlight apparatus. The penlight apparatus is a cylinder, usually oriented horizontally, formed by heating elements 0.6 m (2 ft) long and 0.45 m (1.5 ft) in diameter. In the experiments, the temperature of the cylindrical "shroud" was controlled according to a specified function of time while pairs of cables were monitored, one for thermal and the other for electrical response.

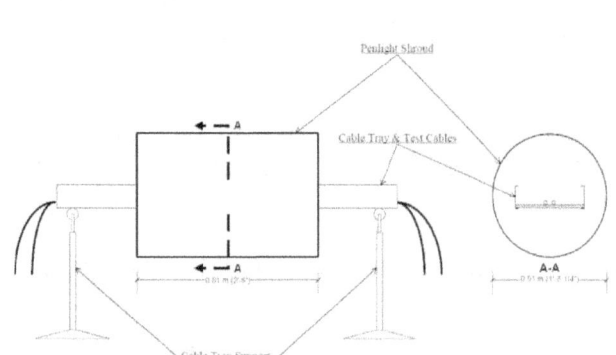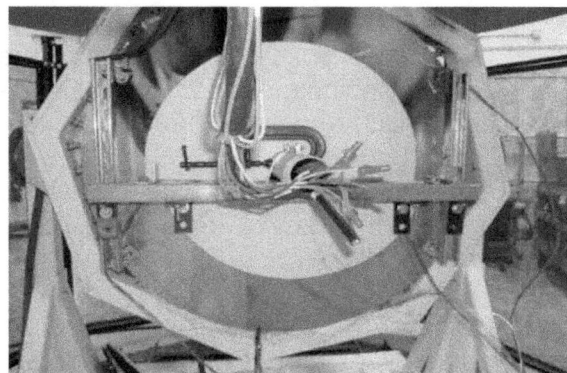

Figure 19-3. Schematic (left) and photograph (right) of the "penlight" test apparatus.

Figure 19-4 shows the results for a 7 conductor cable with XLPE insulation and a CSPE jacket. The penlight apparatus was set to provide an approximate shroud temperature of 475 °C. This cable has jacket thickness of 1.5 mm, an overall diameter of 15 mm, a mass per unit length of 0.410 kg/m, and is a thermoset.

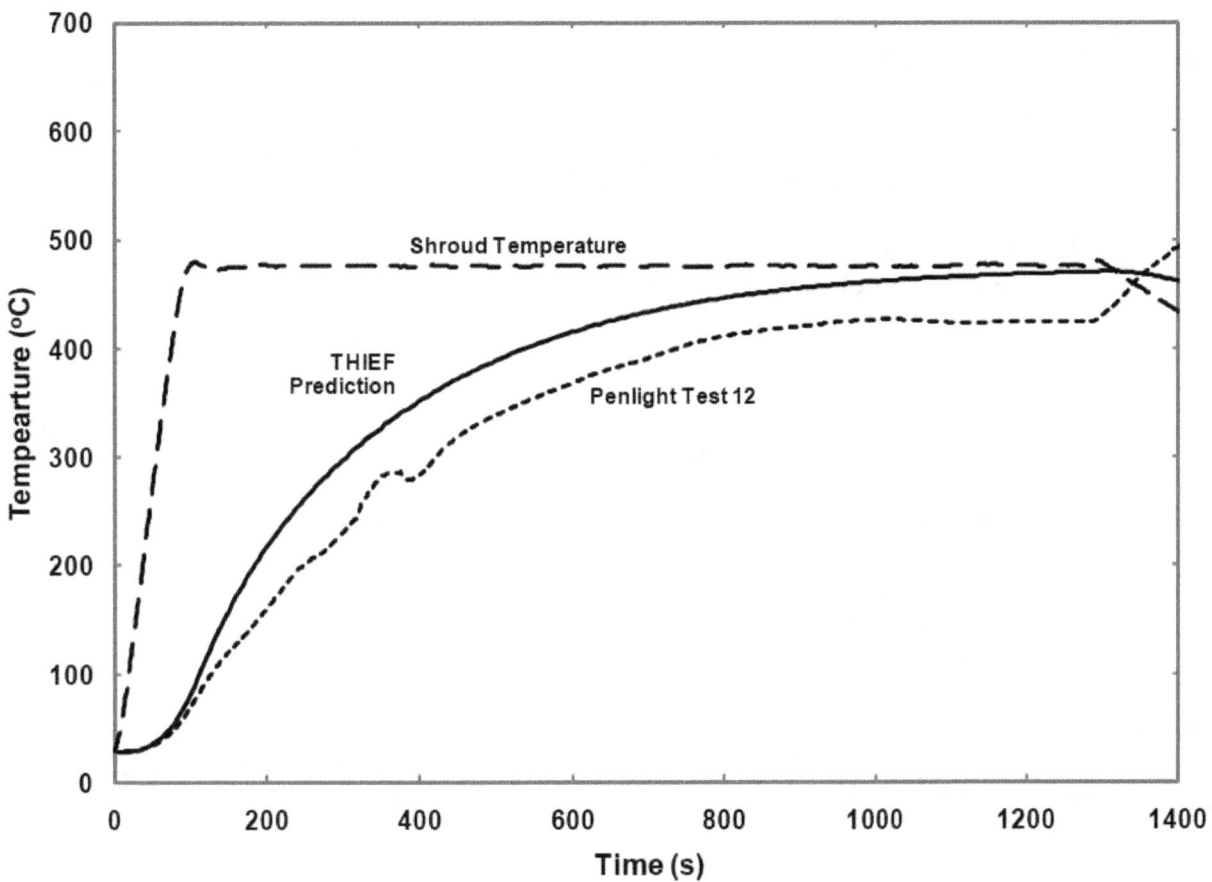

Figure 19-4. Results comparing data (short dash line) for XLPE/CSPE, 7 conductor cable exposed to a shroud temperature (long dash line) of 475 °C and THIEF prediction (solid line).

Solution

Purpose:

    (1)   Determine the time to failure for the cable exposed in the penlight apparatus.

Assumptions:

    (1)   The penlight apparatus exposure is uniform.
    (2)   The cable properties do not change with temperature increase.
    (3)   Failure is indicated when the temperature inside the cable jacket exceeds 400 °C.

Spreadsheet (FDT^s) Information:

Use the following FDT^s:

    (a)   19_THIEF_Thermally_Induced_Electrical_Failure_of_Cables_
           Sup1_SI.xls (click on *THIEF*)

FDT^s Input Parameters:
    -Gas temperature around the cable as a function time (4 points):

Time (s)	Temperature (°C)
0	28
100	475
1300	475
1700	300

    -Type of cable (thermoplastic, thermoset) = Thermoset
    -Cable diameter = 15 mm
    -Cable mass per unit length = 0.410 kg/m
    -Cable jacket thickness =1.5 mm
    -Air Drop

Results*

Time to Cable Failure (minutes)	9.1

*see spreadsheet

The following calculations estimate the time to failure of cables exposed to a specified hot gas layer.

Parameters in YELLOW CELLS are Entered by the User.

Parameters in GREEN CELLS are Automatically Selected from the DROP DOWN MENU or SELECT CABLE BUTTON.

All subsequent output values are calculated by the spreadsheet and based on values specified in the input parameters. This spreadsheet is protected and secure to avoid errors due to a wrong entry in a cell(s). The chapter in the NUREG should be read before an analysis is made.

Project / Inspection Title:	NUREG-1805 Supplement 1 Example 19.11-1

INPUT PARAMETERS

Cable Diameter	15.00	mm	**Do Not Enter Any**
Cable Mass per Unit Length	0.41	kg/m	**Values in the**
Cable Jacket Thickness	1.500	mm	**Green Boxes!**
Ambient Air Temperature	21	°C	
Failure Temperature	400	°C	**They are entered**
Maximum Time	4000	s	**automatically**
Conduit Thickness	0.00	mm	**based on the**
Conduit Outside Diameter	0.00	mm	**cable selection.**
Cable Density	2320.13	kg/m³	
Cable Insulation Type (Thermoplastic or Thermoset)	Thermoset		
Cable Function (Control, Instrumentation or Power)	Instrumentation		
Wire Gauge (AWG)	14		
Number of Conductors	7		
Cable Location	Air Drop		

Select Cable

Click to Select Source for Exposure Gas Temperature Profile:

Natural Ventilation - Method of McCaffrey, Quintiere, Harkleroad (MQH)	**Selecting one of these items will automatically transfer you to the appropriate spreadsheet to calculate the exposure gas temperature profile**
Forced Ventilation - Method of Foote, Pagni, and Alvares (FPA)	
Forced Ventilation - Method of Deal and Beyler	
Room Fire with Closed Door	
Within Fire Plume	
User Defined	

Warning: You MUST Click the Calculate Button Below when Finished Entering or Changing Data!

Calculate

RESULTS

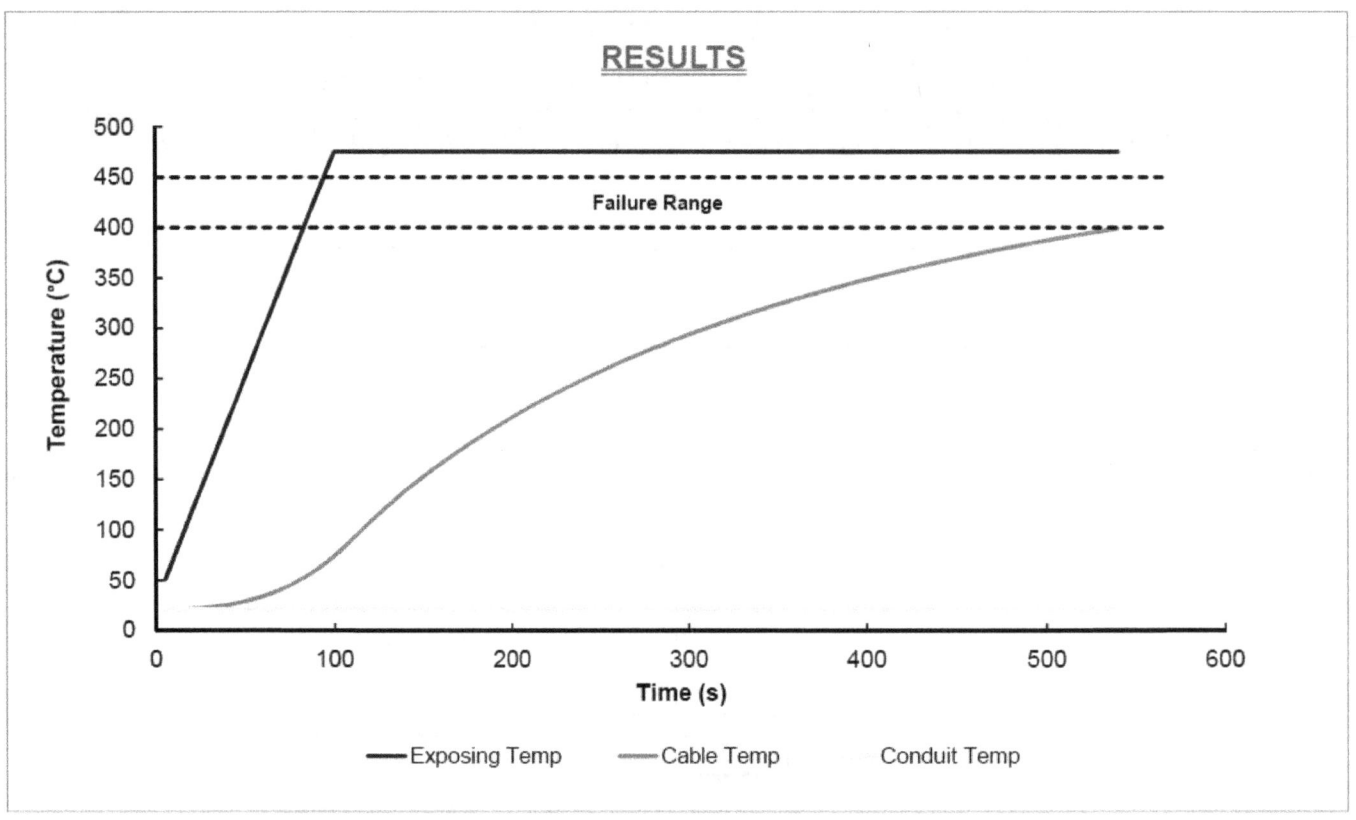

EXPOSURE GAS TEMPERATURE PROFILE		
Time (s)	Gas Temperature (°C)	Gas Temperature (°K)
0	28.00	301.15
100	475.00	748.15
1300	475.00	748.15
1700	300.00	573.15

For User Defined Exposure Gas Temperature Profile:

Enter Time and Temperature Pairs in Table (Temperatures are automatically converted to Kelvin.)

Answer: Cable fails at 9.1 minutes

Example Problem 19.11-2
(Ventilation Differences – Natural vs. Forced)

Problem Statement

Consider a <u>concrete</u> compartment that is 16.40 ft (5.00 m) wide x 16.40 ft (5.00 m) long x 11.48 ft (3.50 m) high (w_c x l_c x h_c), with a simple vent that is 3.28 ft (1.00 m) wide x 6.90 ft (2.10 m) high (w_v x h_v) (open door). Assume the vent is flush with the floor. There is an HVAC system that can be used for smoke exhaust during a fire. The fire is constant with a HRR of 500 kW. A cable tray containing control cables passes through the compartment. The cables are 7 conductor, 10 gauge, power and control cables manufactured by Draka with PVC insulation. Compute the time to cable failure under two scenarios: (a) the HVAC system is off and the room door is open and (b) the door is closed and the HVAC is operating at a rate of 4,000 cfm. The same cable is to be analyzed in both scenarios.

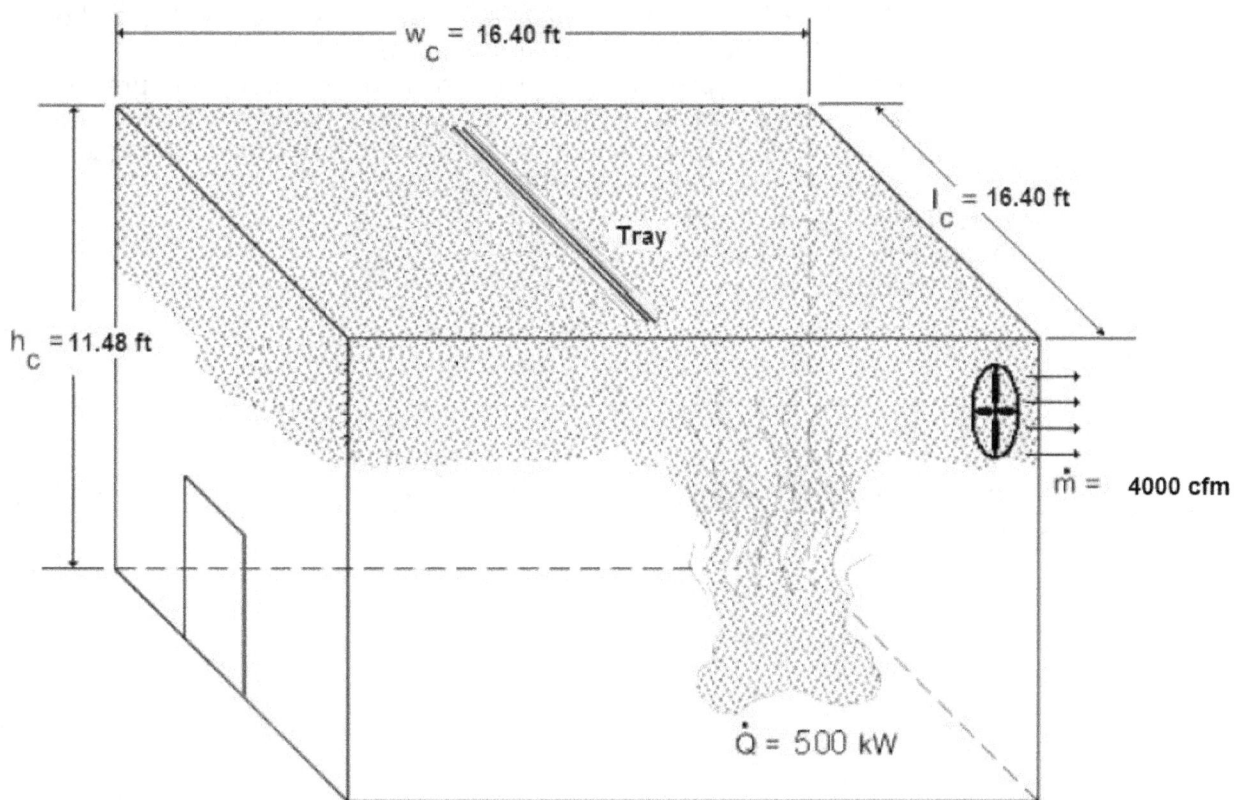

Example 19.11-2. Concrete Compartment with a Door and Forced Ventilation.

Solution

Purpose:

    (1)   Determine the time to failure for the two scenarios and compare the results for natural and forced ventilation.

Assumptions:

    (1)   The cables are surrounded by a uniform temperature hot gas layer.

(2)   The cable properties do not change with temperature increase.

(3)   Failure is indicated when the temperature inside the cable jacket exceeds 390 °F (200 °C).

Spreadsheet (FDT^s) Information:

Use the following FDT^s:

    (a)   02.1_Temperature_NV_Sup1.xls (click on *Temperature_NV*)

    (b)   02.2_Temperature_FV_Sup1.xls (click on *Temperature_FV*)

    (c)   19_THIEF_Thermally_Induced_Electrical_Failure_of_Cables_Sup1.xls (click on *THIEF*)

FDT^s Input Parameters:

-Gas temperature around the cable as a function time:

    (a)   Use 02.1_Temperature_NV_Sup1.xls and the parameters presented above to determine the gas temperature for natural ventilation
On the THIEF spreadsheet, select: Natural Ventilation– Method of McCaffrey, Quintiere, Harkleroad (MQH)

    (b)   Use 02.2_Temperature_FV_Sup1.xls and the parameters presented above to determine the gas temperature for forced ventilation
On the THIEF spreadsheet, select: Forced Ventilation – Method of Foote, Pagni, and Alvares (FPA)

-Press "Select Cable" Button

-Choose appropriate cable from list.

-Press "Calculate" Button

Results*

	Ventilation	
	Natural	Forced
Time to Cable Failure (minutes)	46.2	No Failure

*see spreadsheets

COMPARTMENT WITH THERMALLY THICK/THIN BOUNDARIES

The following calculations estimate the hot gas layer temperature and smoke layer height in enclosure fire.

Parameters in YELLOW CELLS are Entered by the User.

Parameters in GREEN CELLS are Automatically Selected from the DROP DOWN MENU for the Material Selected.

All subsequent output values are calculated by the spreadsheet and based on values specified in the input parameters. This spreadsheet is protected and secure to avoid errors due to a wrong entry in a cell(s). The chapter in the NUREG should be read before an analysis is made.

Project / Inspection Title:	NUREG-1805 Supplement 1 Example 19.11-2a

INPUT PARAMETERS

COMPARTMENT INFORMATION

Compartment Width (w_c)	16.40	ft
Compartment Length (l_c)	16.40	ft
Compartment Height (h_c)	11.48	ft
Vent Width (w_v)	3.28	ft
Vent Height (h_v)	6.90	ft
Top of Vent from Floor (V_T)	6.90	ft
Interior Lining Thickness (δ)	12.00	in

AMBIENT CONDITIONS

Ambient Air Temperature (T_a)	70.00	°F
Specific Heat of Air (c_p)	1.00	kJ/kg-K
Ambient Air Density (ρ_a)	1.20	kg/m³

Note: Ambient Air Density (ρ_a) will automatically correct with Ambient Air Temperature (T_a) Input

THERMAL PROPERTIES OF COMPARTMENT ENCLOSING SURFACES FOR

Interior Lining Thermal Inertia ($k\rho c$)	2.9	(kW/m²-K)²-sec
Interior Lining Thermal Conductivity (k)	0.0016	kW/m-K
Interior Lining Specific Heat (c)	0.75	kJ/kg-K
Interior Lining Density (ρ)	2400	kg/m³

THERMAL PROPERTIES FOR COMMON INTERIOR LINING MATERIALS

Material	$k\rho c$ $(kW/m^2\text{-}K)^2\text{-}sec$	k (kW/m-K)	c (kJ/kg-K)	ρ (kg/m^3)	Select Material
					Concrete ▼
Aluminum (pure)	500	0.206	0.895	2710	Scroll to desired material
Steel (0.5% Carbon)	197	0.054	0.465	7850	Click the selection
Concrete	2.9	0.0016	0.75	2400	
Brick	1.7	0.0008	0.8	2600	
Glass, Plate	1.6	0.00076	0.8	2710	
Brick/Concrete Block	1.2	0.00073	0.84	1900	
Gypsum Board	0.18	0.00017	1.1	960	
Plywood	0.16	0.00012	2.5	540	
Fiber Insulation Board	0.16	0.00053	1.25	240	
Chipboard	0.15	0.00015	1.25	800	
Aerated Concrete	0.12	0.00026	0.96	500	
Plasterboard	0.12	0.00016	0.84	950	
Calcium Silicate Board	0.098	0.00013	1.12	700	
Alumina Silicate Block	0.036	0.00014	1	260	
Glass Fiber Insulation	0.0018	0.000037	0.8	60	
Expanded Polystyrene	0.001	0.000034	1.5	20	
User Specified Value	Enter Value	Enter Value	Enter Value	Enter Value	

Reference: Klote, J., J. Milke, Principles of Smoke Management, 2002, Page 270.

FIRE SPECIFICATIONS

Fire Heat Release Rate (Q)   500.00 kW

Click Here to Calculate this Sheet
and Return to THIEF

Results	Time After Ignition (t)		h_k	ΔT_g	T_g	T_g	T_g
	(min)	(sec)	(kW/m²-K)	(°K)	(°K)	(°C)	(°F)
	0	0.00	-	-	294.11	21.11	70.00
	1	60	0.22	100.57	394.68	121.68	251.03
	2	120	0.16	112.89	407.00	134.00	273.20
	3	180	0.13	120.78	414.89	141.89	287.40
	4	240	0.11	126.71	420.82	147.82	298.08
	5	300	0.10	131.51	425.62	152.62	306.72
	10	600	0.07	147.62	441.73	168.73	335.71
	15	900	0.06	157.94	452.05	179.05	354.29
	20	1200	0.05	165.70	459.81	186.81	368.25
	25	1500	0.04	171.97	466.09	193.09	379.55
	30	1800	0.04	177.28	471.39	198.39	389.11
	35	2100	0.04	181.89	476.01	203.01	397.41
	40	2400	0.03	185.99	480.10	207.10	404.78
	45	2700	0.03	189.68	483.79	210.79	411.42
	50	3000	0.03	193.04	487.15	214.15	417.46
	55	3300	0.03	196.13	490.24	217.24	423.03
	60	3600	0.03	198.99	493.10	220.10	428.18

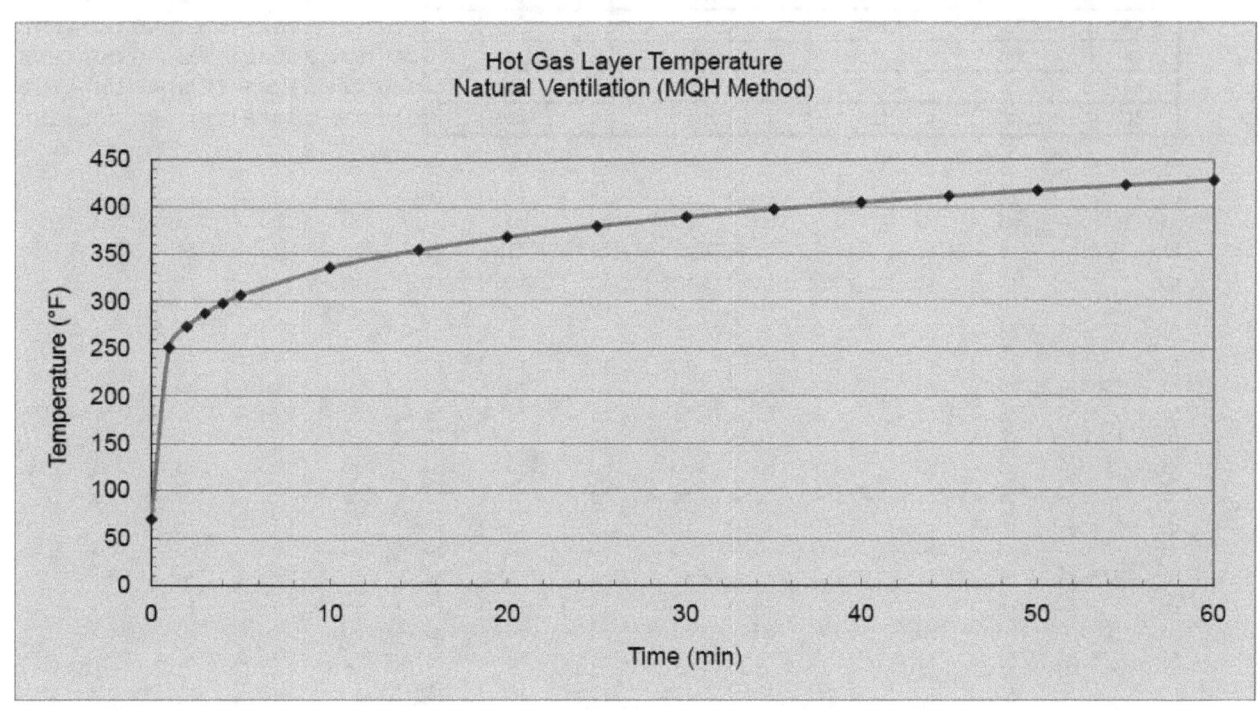

Results	Caution! The smoke layer height is a conservative estimate and is only intended to provide an indication where the hot gas layer is located. Calculated smoke layer height below the vent height are not creditable since the calculation is not accounting for the smoke exiting the vent.

Time (min)	ρ_g (kg/m^3)	Constant (k) (kW/m-K)	Smoke Layer Height z (m)	Smoke Layer Height z (ft)	
0	1.20	0.063	3.50	11.48	
1	0.89	0.085	2.10	6.90	CAUTION: SMOKE IS EXITING OUT VENT
2	0.87	0.088	2.10	6.90	CAUTION: SMOKE IS EXITING OUT VENT
3	0.85	0.089	2.10	6.90	CAUTION: SMOKE IS EXITING OUT VENT
4	0.84	0.091	2.10	6.90	CAUTION: SMOKE IS EXITING OUT VENT
5	0.83	0.092	2.10	6.90	CAUTION: SMOKE IS EXITING OUT VENT
10	0.80	0.095	2.10	6.90	CAUTION: SMOKE IS EXITING OUT VENT
15	0.78	0.097	2.10	6.90	CAUTION: SMOKE IS EXITING OUT VENT
20	0.77	0.099	2.10	6.90	CAUTION: SMOKE IS EXITING OUT VENT
25	0.76	0.100	2.10	6.90	CAUTION: SMOKE IS EXITING OUT VENT
30	0.75	0.101	2.10	6.90	CAUTION: SMOKE IS EXITING OUT VENT
35	0.74	0.102	2.10	6.90	CAUTION: SMOKE IS EXITING OUT VENT
40	0.74	0.103	2.10	6.90	CAUTION: SMOKE IS EXITING OUT VENT
45	0.73	0.104	2.10	6.90	CAUTION: SMOKE IS EXITING OUT VENT
50	0.72	0.105	2.10	6.90	CAUTION: SMOKE IS EXITING OUT VENT
55	0.72	0.106	2.10	6.90	CAUTION: SMOKE IS EXITING OUT VENT
60	0.72	0.106	2.10	6.90	CAUTION: SMOKE IS EXITING OUT VENT

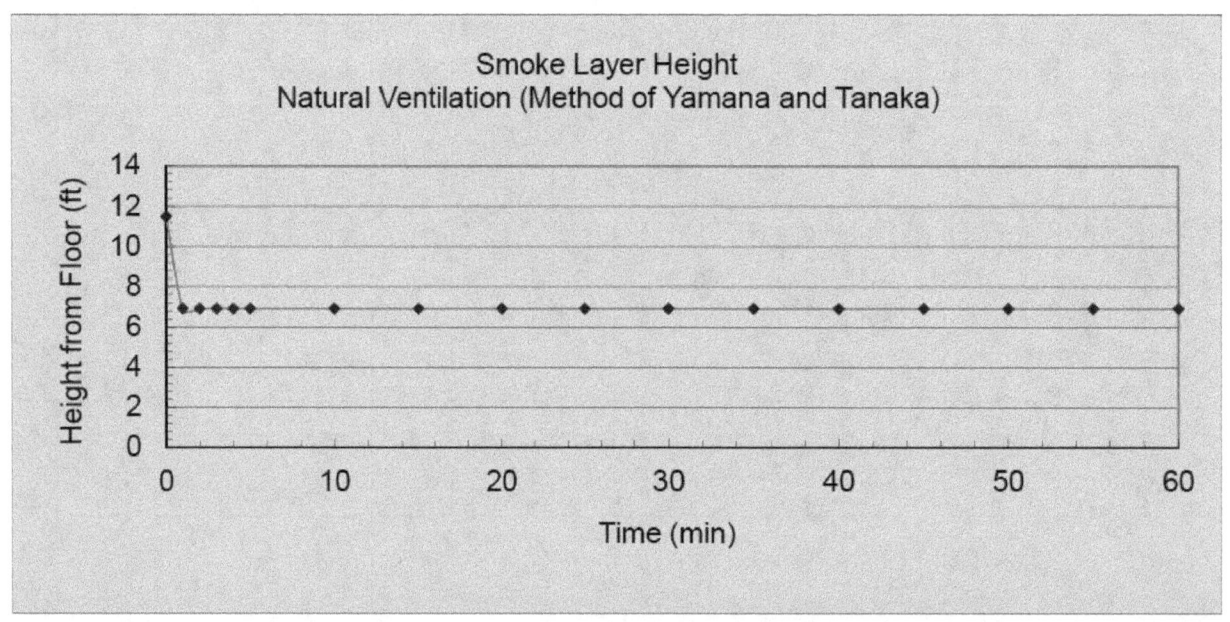

Smoke Layer Height
Natural Ventilation (Method of Yamana and Tanaka)

NOTE:

The above calculations are based on principles developed in the SFPE Handbook of Fire Protection Engineering, 3rd Edition, 2002. Calculations are based on certain assumptions and have inherent limitations. The results of such calculations may or may not have reasonable predictive capabilities for a given situation and should only be interpreted by an informed user. Although each calculation in the spreadsheet has been verified with the results of hand calculation, there is no absolute guarantee of the accuracy of these calculations. Any questions, comments, concerns, and suggestions, or to report an error(s) in the spreadsheet, please send an email to David.Stroup@nrc.gov and Naeem.Iqbal@nrc.gov or MarkHenry.Salley@nrc.gov.

The following calculations estimate the time to failure of cables exposed to a specified hot gas layer.

Parameters in YELLOW CELLS are Entered by the User.

Parameters in GREEN CELLS are Automatically Selected from the DROP DOWN MENU or SELECT CABLE BUTTON.

All subsequent output values are calculated by the spreadsheet and based on values specified in the input parameters. This spreadsheet is protected and secure to avoid errors due to a wrong entry in a cell(s). The chapter in the NUREG should be read before an analysis is made.

Project / Inspection Title:	NUREG-1805 Supplement 1 Example 19.11-2a

INPUT PARAMETERS

Cable Diameter	0.65	in
Cable Mass per Unit Length	0.36	b/ft
Cable Jacket Thickness	0.060	in
Ambient Air Temperature	70	°F
Failure Temperature	392	°F
Maximum Time	4000	s
Conduit Thickness	0.00	in
Conduit Outside Diameter	0.00	in
Cable Density	2516.39	kg/m³
Cable Insulation Type (Thermoplastic or Thermoset)	Thermoplastic	
Cable Function (Control, Instrumentation or Power)	Control	
Wire Gauge (AWG)	10	
Number of Conductors	7	
Cable Location	Cable Tray	

Do Not Enter Any Values in the Green Boxes!

They are entered automatically based on the cable selection.

[Select Cable]

Click to Select Source for Exposure Gas Temperature Profile:

Natural Ventilation - Method of McCaffrey, Quintiere, Harkleroad (MQH)
Forced Ventilation - Method of Foote, Pagni, and Alvares (FPA)
Forced Ventilation - Method of Deal and Beyler
Room Fire with Closed Door
Within Fire Plume
User Defined

Selecting one of these items will automatically transfer you to the appropriate spreadsheet to calculate the exposure gas temperature profile

Warning: You MUST Click the Calculate Button Below when Finished Entering or Changing Data!

[Calculate]

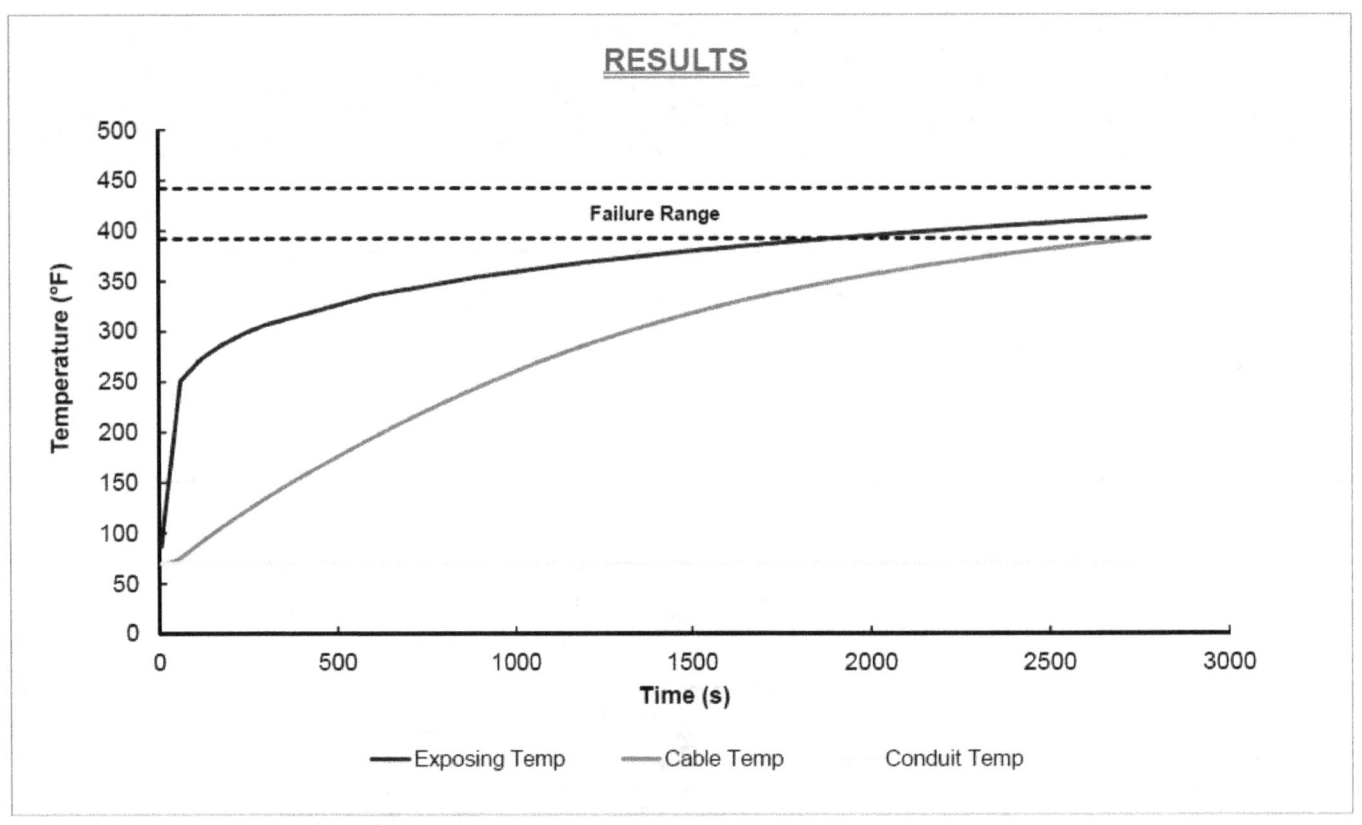

RESULTS

EXPOSURE GAS TEMPERATURE PROFILE		
Time (s)	Gas Temperature (°F)	Gas Temperature (°K)
0	70.00	294.26
60	251.03	394.83
120	273.20	407.15
180	287.40	415.04
240	298.08	420.97
300	306.72	425.77
600	335.71	441.88
900	354.29	452.20
1200	368.25	459.96
1500	379.55	466.24
1800	389.11	471.54
2100	397.41	476.16
2400	404.78	480.25
2700	411.42	483.94
3000	417.46	487.30
3300	423.03	490.39
3600	428.18	493.25

Answer: Cable fails at 46.2 minutes

CHAPTER 2. PREDICTING HOT GAS LAYER TEMPERATURE IN A ROOM FIRE WITH FORCED VENTILATION

Version 1805.1 (English Units)

COMPARTMENT WITH THERMALLY THICK/THIN BOUNDARIES

The following calculations estimate the hot gas layer temperature and smoke layer height in enclosure fire.

Parameters in YELLOW CELLS are Entered by the User.

Parameters in GREEN CELLS are Automatically Selected from the DROP DOWN MENU for the Material Selected.

All subsequent output values are calculated by the spreadsheet and based on values specified in the input parameters. This spreadsheet is protected and secure to avoid errors due to a wrong entry in a cell(s). The chapter in the NUREG should be read before an analysis is made.

Project / Inspection Title:	NUREG-1805 Supplement 1 Example 19-11.2b

INPUT PARAMETERS

COMPARTMENT INFORMATION

Compartment Width (w_c)	16.40	ft
Compartment Length (l_c)	16.40	ft
Compartment Height (h_c)	11.48	ft
Interior Lining Thickness (δ)	12.00	in

AMBIENT CONDITIONS

Ambient Air Temperature (T_a)	70.00	°F
Specific Heat of Air (c_p)	1.00	kJ/kg-K
Ambient Air Density (ρ_a)	1.20	kg/m^3

Note: Ambient Air Density (ρ_a) will automatically correct with Ambient Air Temperature (T_a) Input

THERMAL PROPERTIES OF COMPARTMENT ENCLOSING SURFACES FOR

Interior Lining Thermal Inertia ($k\rho c$)	2.9	(kW/m^2-K)^2-sec
Interior Lining Thermal Conductivity (k)	0.0016	kW/m-K
Interior Lining Specific Heat (c)	0.75	kJ/kg-K
Interior Lining Density (ρ)	2400	kg/m^3

CHAPTER 2. PREDICTING HOT GAS LAYER TEMPERATURE
IN A ROOM FIRE
WITH FORCED VENTILATION

**Version 1805.1
(English Units)**

THERMAL PROPERTIES FOR COMMON INTERIOR LINING MATERIALS

Material	kρc (kW/m²-K)²-sec	k (kW/m-K)	c (kJ/kg-K)	ρ (kg/m³)	Select Material
Aluminum (pure)	500	0.206	0.895	2710	Concrete ▾
Steel (0.5% Carbon)	197	0.054	0.465	7850	**Scroll to desired material**
Concrete	2.9	0.0016	0.75	2400	**Click on selection**
Brick	1.7	0.0008	0.8	2600	
Glass, Plate	1.6	0.00076	0.8	2710	
Brick/Concrete Block	1.2	0.00073	0.84	1900	
Gypsum Board	0.18	0.00017	1.1	960	
Plywood	0.16	0.00012	2.5	540	
F ber Insulation Board	0.16	0.00053	1.25	240	
Chipboard	0.15	0.00015	1.25	800	
Aerated Concrete	0.12	0.00026	0.96	500	
Plasterboard	0.12	0.00016	0.84	950	
Calcium Silicate Board	0.098	0.00013	1.12	700	
Alumina Silicate Block	0.036	0.00014	1	260	
Glass Fiber Insulation	0.0018	0.000037	0.8	60	
Expanded Polystyrene	0.001	0.000034	1.5	20	
User Specified Value	Enter Value	Enter Value	Enter Value	Enter Value	

Reference: Klote, J., J. Milke, Principles of Smoke Management, 2002 Page 270.

COMPARTMENT MASS VENTILATION FLOW RATE

Forced Ventilation Flow Rate (m)   4000.00 cfm

FIRE SPECIFICATIONS

Fire Heat Release Rate (Q)   500.00 kW

**Click Here to Calculate this Sheet
and Return to THIEF**

Compartment Hot Gas Layer Temperature With Forced Ventilation

$$\Delta T_g/T_a = 0.63(Q/mc_pT_a)^{0.72}(h_kA_T/mc_p)^{-0.36}$$

$$\Delta T_g = T_g - T_a$$

$$T_g = \Delta T_g + T_a$$

Results	Time After Ignition (t)		h_k	$\Delta T_g/T_a$	ΔT_g	T_g	T_g	T_g
	(min)	(sec)	(kW/m²-K)		(°K)	(°K)	(°C)	(°F)
	0	0	-	-	-	294.11	21.11	70.00
	1	60	0.22	0.21	62.28	356.39	83.39	182.10
	2	120	0.16	0.24	70.55	364.66	91.66	196.99
	3	180	0.13	0.26	75.89	370.00	97.00	206.61
	4	240	0.11	0.27	79.93	374.04	101.04	213.87
	5	300	0.10	0.28	83.20	377.31	104.31	219.76
	10	600	0.07	0.32	94.26	388.37	115.37	239.66
	15	900	0.06	0.34	101.39	395.50	122.50	252.51
	20	1200	0.05	0.36	106.78	400.89	127.89	262.21
	25	1500	0.04	0.38	111.16	405.27	132.27	270.09
	30	1800	0.04	0.39	114.87	408.98	135.98	276.76
	35	2100	0.04	0.40	118.10	412.21	139.21	282.58
	40	2400	0.03	0.41	120.97	415.08	142.08	287.75
	45	2700	0.03	0.42	123.56	417.68	144.68	292.42
	50	3000	0.03	0.43	125.93	420.04	147.04	296.67
	55	3300	0.03	0.44	128.11	422.22	149.22	300.60
	60	3600	0.03	0.44	130.13	424.24	151.24	304.24

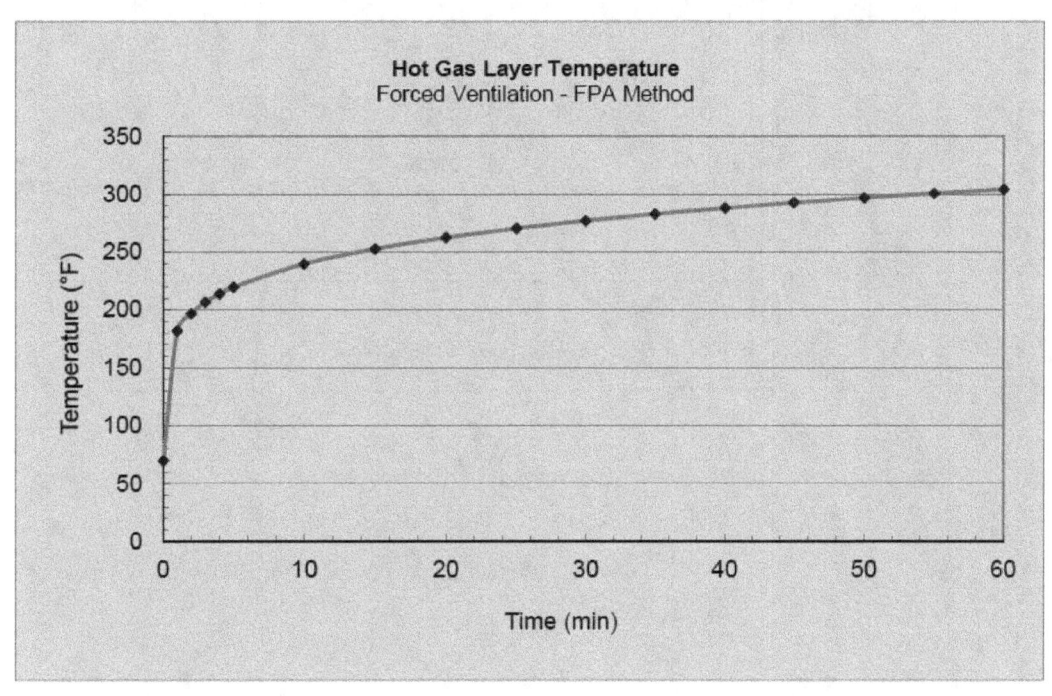

Hot Gas Layer Temperature
Forced Ventilation - FPA Method

Results	Time After Ignition (t)		h_k	ΔT_g	T_g	T_g	T_g
	(min)	(sec)	$(kW/m^2\text{-}K)$	(°K)	(°K)	(°C)	(°F)
	0	0	-	-	294.11	21.11	70.00
	1	60	0.09	39.02	333.13	60.13	140.24
	2	120	0.06	51.42	345.53	72.53	162.56
	3	180	0.05	59.84	353.95	80.95	177.72
	4	240	0.04	66.32	360.43	87.43	189.37
	5	300	0.04	71.61	365.72	92.72	198.89
	10	600	0.03	89.27	383.38	110.38	230.68
	15	900	0.02	100.22	394.33	121.33	250.39
	20	1200	0.02	108.13	402.24	129.24	264.63
	25	1500	0.02	114.28	408.39	135.39	275.70
	30	1800	0.02	119.29	413.40	140.40	284.72
	35	2100	0.01	123.50	417.61	144.61	292.30
	40	2400	0.01	127.11	421.23	148.23	298.81
	45	2700	0.01	130.27	424.38	151.38	304.49
	50	3000	0.01	133.07	427.18	154.18	309.53
	55	3300	0.01	135.58	429.69	156.69	314.04
	60	3600	0.01	137.84	431.95	158.95	318.11

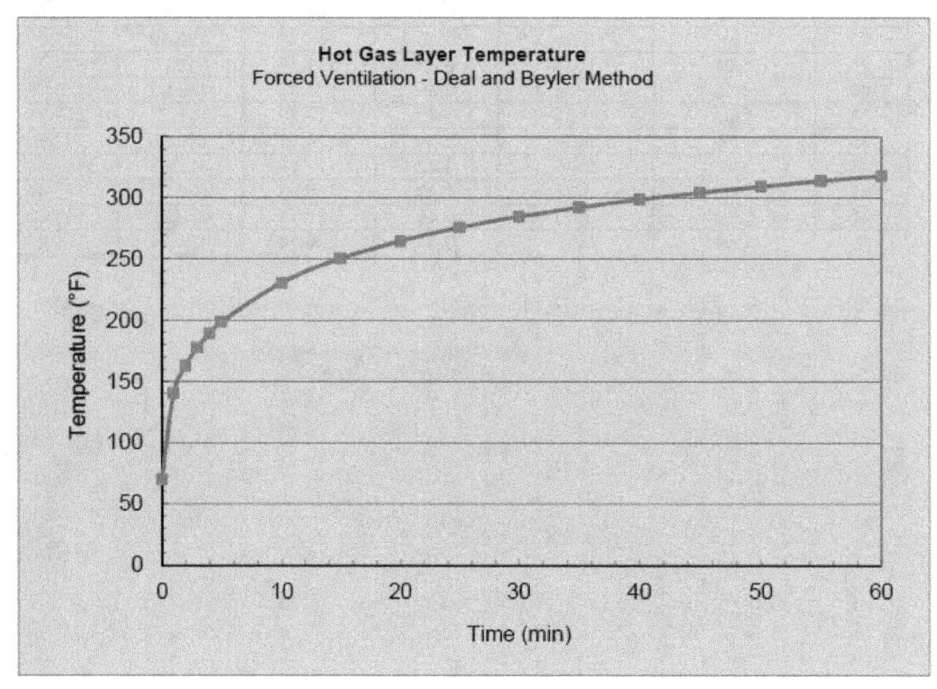

Summary of Results

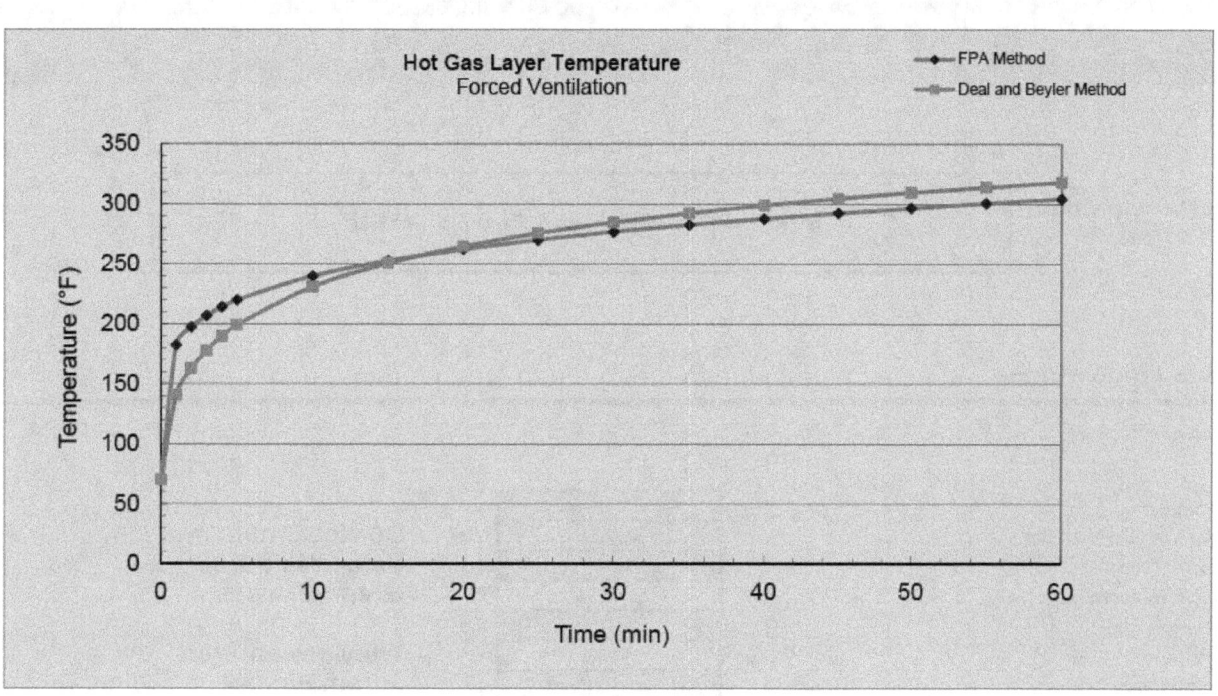

NOTE:
The above calculations are based on principles developed in the SFPE Handbook of Fire Protection Engineering, 2nd Edition, 1995. Calculations are based on certain assumptions and have inherent limitations. The results of such calculations may or may not have reasonable predictive capabilities for a given situation, and should only be interpreted by an informed user. Although each calculation in the spreadsheet has been verified with the results of hand calculation, there is no absolute guarantee of the accuracy of these calculations. Any questions, comments, concerns, and suggestions, or to report an error(s) in the spreadsheets, please send an email to David.Stroup@nrc.gov and Naeem.Iqbal@nrc.gov or MarkHenry.Salley@nrc.gov.

Prepared by: [   ]   Date: [   ]   Organization: [   ]

Checked by: [   ]   Date: [   ]   Organization: [   ]

Additional Information:

[   ]

The following calculations estimate the time to failure of cables exposed to a specified hot gas layer.

Parameters in YELLOW CELLS are Entered by the User.

Parameters in GREEN CELLS are Automatically Selected from the DROP DOWN MENU or SELECT CABLE BUTTON.

All subsequent output values are calculated by the spreadsheet and based on values specified in the input parameters. This spreadsheet is protected and secure to avoid errors due to a wrong entry in a cell(s). The chapter in the NUREG should be read before an analysis is made.

	Project / Inspection Title:	NUREG-1805 Supplement 1 Example 19.11-2b

INPUT PARAMETERS

Cable Diameter	0.65	in
Cable Mass per Unit Length	0.36	b/ft
Cable Jacket Thickness	0.060	in
Ambient Air Temperature	70	°F
Failure Temperature	392	°F
Maximum Time	4000	s
Conduit Thickness	0.00	in
Conduit Outside Diameter	0.00	in
Cable Density	2516.39	kg/m³
Cable Insulation Type (Thermoplastic or Thermoset)	Thermoplastic	
Cable Function (Control, Instrumentation or Power)	Control	
Wire Gauge (AWG)	10	
Number of Conductors	7	
Cable Location	Cable Tray	

Do Not Enter Any Values in the Green Boxes!

They are entered automatically based on the cable selection.

Select Cable

Click to Select Source for Exposure Gas Temperature Profile:

Natural Ventilation - Method of McCaffrey, Quintiere, Harkleroad (MQH)
Forced Ventilation - Method of Foote, Pagni, and Alvares (FPA)
Forced Ventilation - Method of Deal and Beyler
Room Fire with Closed Door
Within Fire Plume
User Defined

Selecting one of these items will automatically transfer you to the appropriate spreadsheet to calculate the exposure gas temperature profile

Warning: You MUST Click the Calculate Button Below when Finished Entering or Changing Data!

Calculate

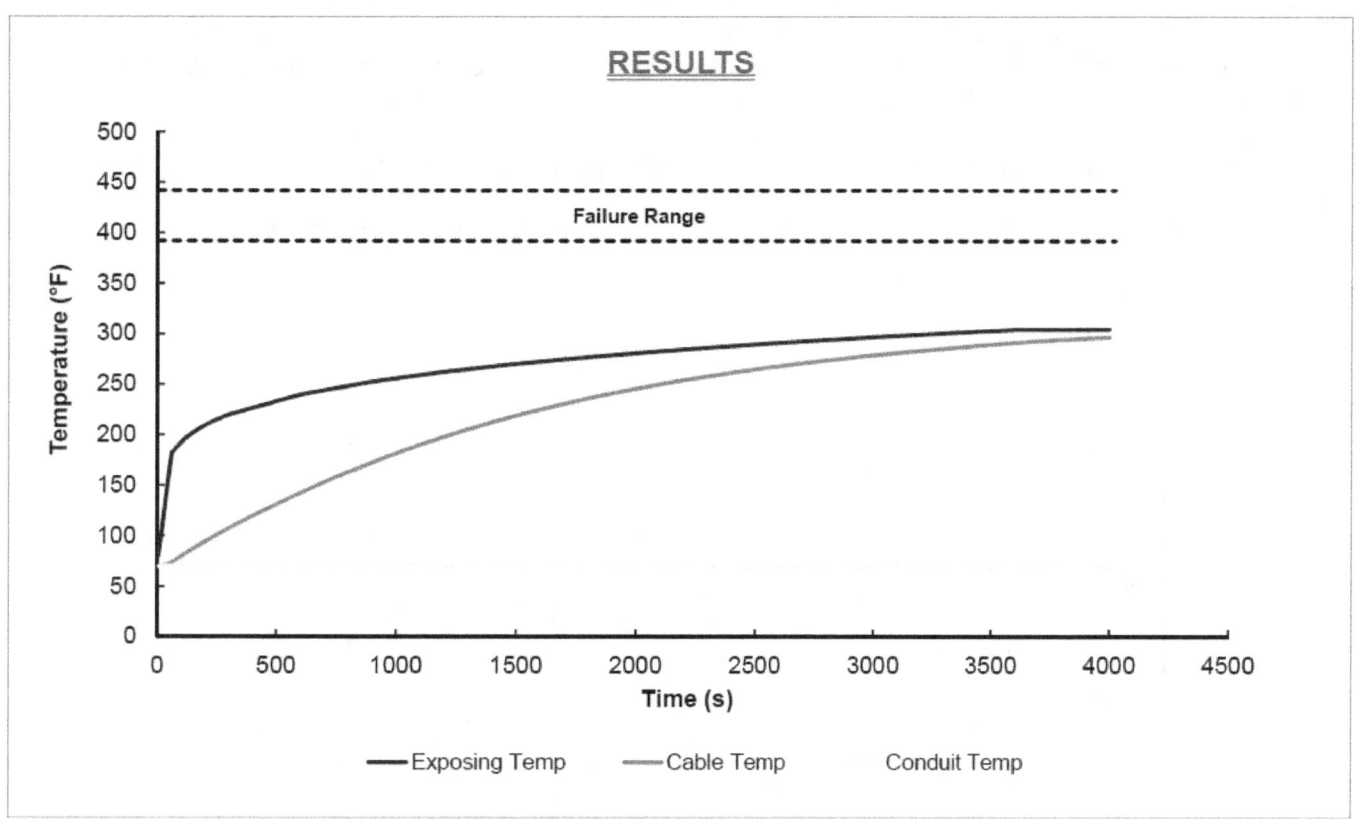

RESULTS

EXPOSURE GAS TEMPERATURE PROFILE		
Time (s)	Gas Temperature (°F)	Gas Temperature (°K)
0	70.00	294.26
60	182.10	356.54
120	196.99	364.81
180	206.61	370.15
240	213.87	374.19
300	219.76	377.46
600	239.66	388.52
900	252.51	395.65
1200	262.21	401.04
1500	270.09	405.42
1800	276.76	409.13
2100	282.58	412.36
2400	287.75	415.23
2700	292.42	417.83
3000	296.67	420.19
3300	300.60	422.37
3600	304.24	424.39

Answer: Cable does not reach failure temperature in 4000.2 seconds

Example Problem 19.11-3
(Cable Location – Cable Tray vs. Conduit - Natural Ventilation)

Problem Statement
Consider a <u>concrete</u> compartment that is 16.40 ft (5.00 m) wide x 32.81 ft (10.00 m) long x 9.84 ft (3.00 m) high (w_c x l_c x h_c), with a simple vent that is 3.28 ft (1.00 m) wide x 6.90 ft (2.10 m) high (w_v x h_v) (open door). Assume the vent is flush with the floor. The fire is constant with a HRR of 575 kW and no HVAC is located in this room. Compute the time to cable failure under two scenarios: (a) the cable is located in a cable tray and (b) the cable is located in a 3-inch (78 mm) rigid conduit. The same cable is to be analyzed in both scenarios, and the identification print located on the exterior of the cable jacket reads the following:

GENERAL CABLE® 20/10 CONTROL CABLE 7/C 12AWG PE/PVC 75°C 600V

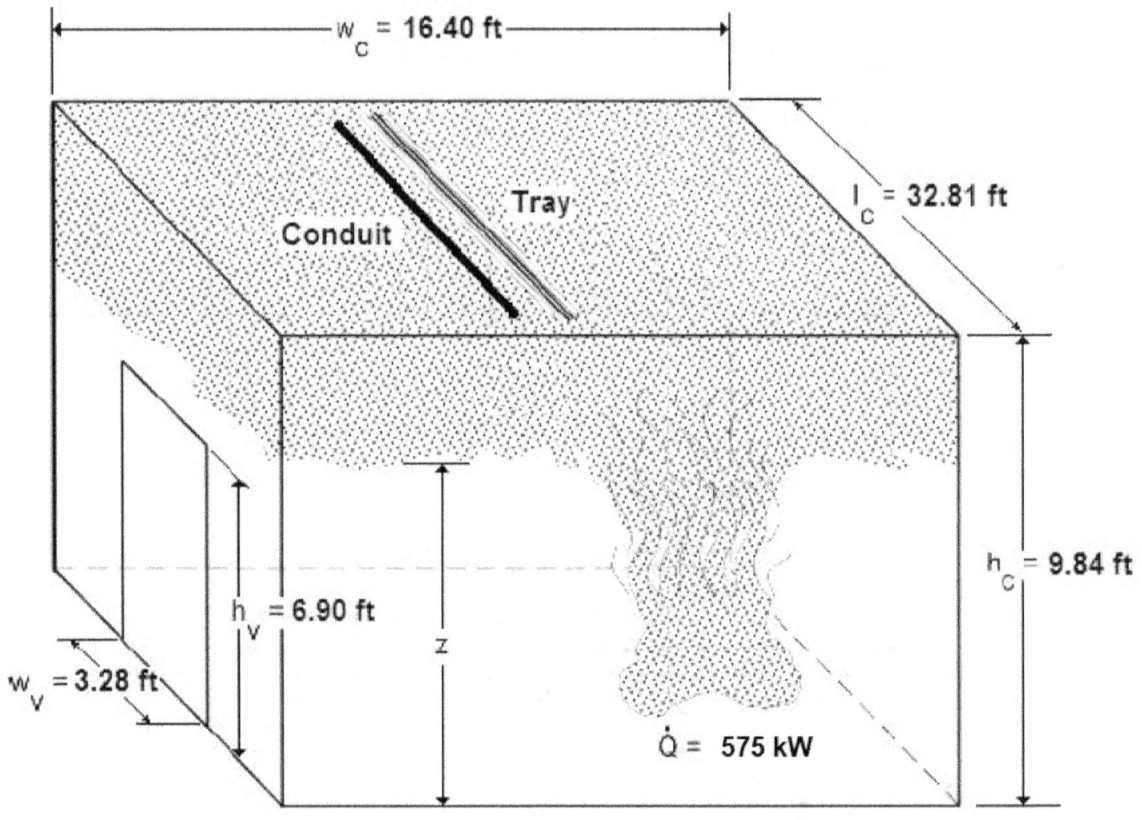

Example 19.11-3. Concrete Compartment with Natural Ventilation.

Solution

Purpose:

    (1)   Determine the time to failure and compare the results of an electrical cable located in a cable tray and one located in a rigid conduit.

Assumptions:

    (1)   The cables are surrounded by a uniform temperature hot gas layer.
    (2)   The cable properties do not change with temperature increase.
    (3)   Failure is indicated when the temperature inside the cable jacket exceeds 392°F (200°C) (thermoplastic cable).

Spreadsheet (FDT^s) Information:

Use the following FDT^s:

    (a)   02.1_Temperature_NV_Sup1.xls (click on *Temperature_NV*)
    (b)   19_THIEF_Thermally_Induced_Electrical_Failure_of_Cables_Sup1.xls (click on *THIEF*)

FDT^s Input Parameters
    -Gas temperature around the cable as a function time:
        o   Use 02.1_Temperature_NV_Sup1.xls and the parameters presented above to determine the gas temperature (same for both scenarios)
        o   On the THIEF spreadsheet, select: Natural Ventilation – Method of McCaffrey, Quintiere, Harkleroad (MQH)
    -Press "Select Cable" Button
    -Choose appropriate cable from list (i.e., Control – 12 AWG – 7 conductor – General Cable – 20/10 – PE – PVC)
    - Choose appropriate "Cable Location"
        (a) "Cable Tray"
        (b) "Conduit – Rigid" & 3" for Conduit outside diameter
    -Press "Calculate" Button

Results*

Cable Tray – Time to Cable Failure (minutes)	54.6
Cable Conduit – Time to Cable Failure (minutes)	No failure

*see spreadsheets

This example illustrates the additional thermal protection provided by the conduit. The conduit has a mass that must first be heated before the cable can be exposed to elevated temperature. The conduit's thermal inertia provides the additional time (protection) to cable failure.

COMPARTMENT WITH THERMALLY THICK/THIN BOUNDARIES

The following calculations estimate the hot gas layer temperature and smoke layer height in enclosure fire.
Parameters in YELLOW CELLS are Entered by the User.
Parameters in GREEN CELLS are Automatically Selected from the DROP DOWN MENU for the Material Selected.
All subsequent output values are calculated by the spreadsheet and based on values specified in the input parameters. This spreadsheet is protected and secure to avoid errors due to a wrong entry in a cell(s). The chapter in the NUREG should be read before an analysis is made.

Project / Inspection Title:	NUREG-1805 Supplement 1 Example 19-11.3a

INPUT PARAMETERS

COMPARTMENT INFORMATION

Compartment Width (w_c)	16.40	ft
Compartment Length (l_c)	32.81	ft
Compartment Height (h_c)	9.84	ft
Vent Width (w_v)	3.28	ft
Vent Height (h_v)	6.90	ft
Top of Vent from Floor (V_T)	6.90	ft
Interior Lining Thickness (δ)	12.00	in

AMBIENT CONDITIONS

Ambient Air Temperature (T_a)	70.00	°F
Specific Heat of Air (c_p)	1.00	kJ/kg-K
Ambient Air Density (ρ_a)	1.20	kg/m^3

Note: Ambient Air Density (ρ_a) will automatically correct with Ambient Air Temperature (T_a) Input

THERMAL PROPERTIES OF COMPARTMENT ENCLOSING SURFACES FOR

Interior Lining Thermal Inertia ($k\rho c$)	2.9	(kW/m^2-K)^2-sec
Interior Lining Thermal Conductivity (k)	0.0016	kW/m-K
Interior Lining Specific Heat (c)	0.75	kJ/kg-K
Interior Lining Density (ρ)	2400	kg/m^3

THERMAL PROPERTIES FOR COMMON INTERIOR LINING MATERIALS

Material	$k\rho c$ $(kW/m^2\text{-}K)^2\text{-sec}$	k $(kW/m\text{-}K)$	c $(kJ/kg\text{-}K)$	ρ (kg/m^3)	Select Material
					Concrete ▾
Aluminum (pure)	500	0.206	0.895	2710	Scroll to desired material
Steel (0.5% Carbon)	197	0.054	0.465	7850	Click the selection
Concrete	2.9	0.0016	0.75	2400	
Brick	1.7	0.0008	0.8	2600	
Glass, Plate	1.6	0.00076	0.8	2710	
Brick/Concrete Block	1.2	0.00073	0.84	1900	
Gypsum Board	0.18	0.00017	1.1	960	
Plywood	0.16	0.00012	2.5	540	
Fiber Insulation Board	0.16	0.00053	1.25	240	
Chipboard	0.15	0.00015	1.25	800	
Aerated Concrete	0.12	0.00026	0.96	500	
Plasterboard	0.12	0.00016	0.84	950	
Calcium Silicate Board	0.098	0.00013	1.12	700	
Alumina Silicate Block	0.036	0.00014	1	260	
Glass Fiber Insulation	0.0018	0.000037	0.8	60	
Expanded Polystyrene	0.001	0.000034	1.5	20	
User Specified Value	Enter Value	Enter Value	Enter Value	Enter Value	

Reference: Klote, J., J. Milke, Principles of Smoke Management, 2002, Page 270.

FIRE SPECIFICATIONS

Fire Heat Release Rate (Q)   575.00 kW

**Click Here to Calculate this Sheet
and Return to THIEF**

Results

Time After Ignition (t)		h_k	ΔT_g	T_g	T_g	T_g
(min)	(sec)	(kW/m²-K)	(°K)	(°K)	(°C)	(°F)
0	0.00	-	-	294.11	21.11	70.00
1	60	0.22	94.50	388.61	115.61	240.10
2	120	0.16	106.07	400.18	127.18	260.93
3	180	0.13	113.49	407.60	134.60	274.28
4	240	0.11	119.06	413.17	140.17	284.31
5	300	0.10	123.57	417.68	144.68	292.43
10	600	0.07	138.71	432.82	159.82	319.67
15	900	0.06	148.40	442.52	169.52	337.13
20	1200	0.05	155.69	449.80	176.80	350.25
25	1500	0.04	161.59	455.70	182.70	360.87
30	1800	0.04	166.58	460.69	187.69	369.84
35	2100	0.04	170.91	465.02	192.02	377.64
40	2400	0.03	174.76	468.87	195.87	384.57
45	2700	0.03	178.22	472.33	199.33	390.80
50	3000	0.03	181.38	475.49	202.49	396.49
55	3300	0.03	184.29	478.40	205.40	401.71
60	3600	0.03	186.98	481.09	208.09	406.56

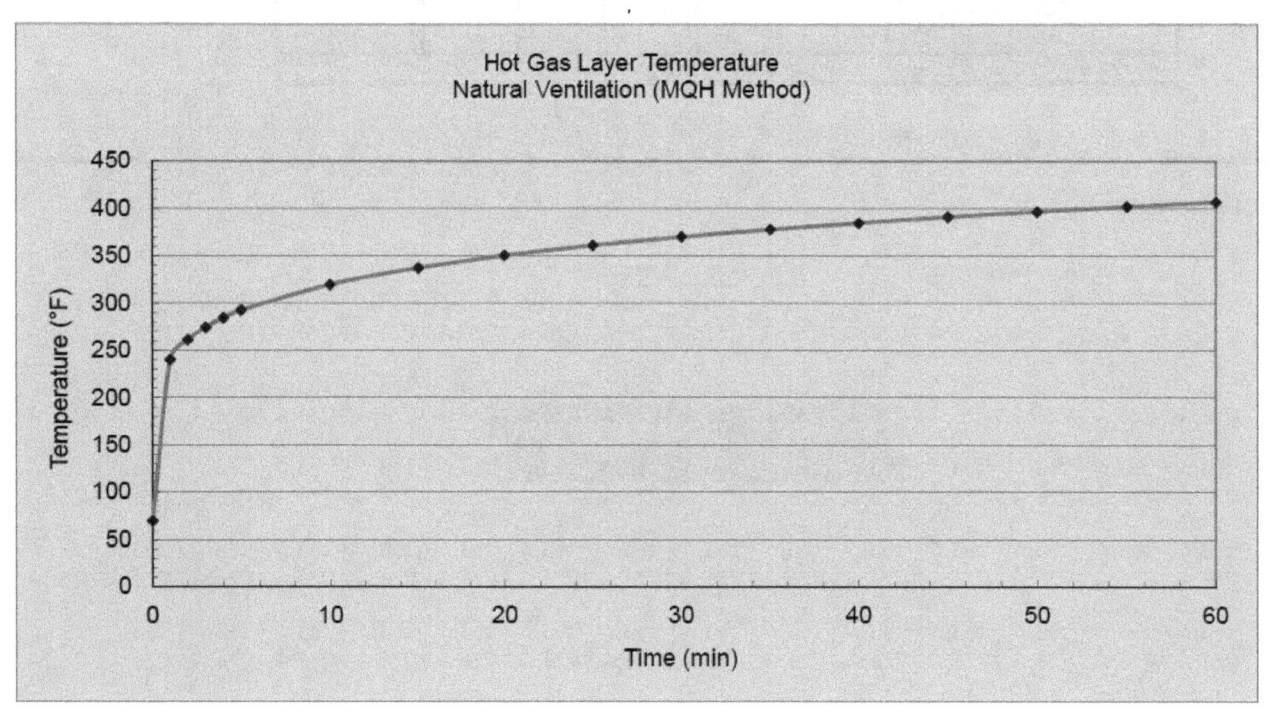

Hot Gas Layer Temperature
Natural Ventilation (MQH Method)

Results	Caution! The smoke layer height is a conservative estimate and is only intended to provide an indication where the hot gas layer is located. Calculated smoke layer height below the vent height are not creditable since the calculation is not accounting for the smoke exiting the vent.

Time (min)	ρ_g (kg/m³)	Constant (k) (kW/m-K)	Smoke Layer Height z (m)	Smoke Layer Height z (ft)	
0	1.20	0.063	3.00	9.84	
1	0.91	0.084	2.10	6.90	CAUTION: SMOKE IS EXITING OUT VENT
2	0.88	0.086	2.10	6.90	CAUTION: SMOKE IS EXITING OUT VENT
3	0.87	0.088	2.10	6.90	CAUTION: SMOKE IS EXITING OUT VENT
4	0.85	0.089	2.10	6.90	CAUTION: SMOKE IS EXITING OUT VENT
5	0.85	0.090	2.10	6.90	CAUTION: SMOKE IS EXITING OUT VENT
10	0.82	0.093	2.10	6.90	CAUTION: SMOKE IS EXITING OUT VENT
15	0.80	0.095	2.10	6.90	CAUTION: SMOKE IS EXITING OUT VENT
20	0.78	0.097	2.10	6.90	CAUTION: SMOKE IS EXITING OUT VENT
25	0.77	0.098	2.10	6.90	CAUTION: SMOKE IS EXITING OUT VENT
30	0.77	0.099	2.10	6.90	CAUTION: SMOKE IS EXITING OUT VENT
35	0.76	0.100	2.10	6.90	CAUTION: SMOKE IS EXITING OUT VENT
40	0.75	0.101	2.10	6.90	CAUTION: SMOKE IS EXITING OUT VENT
45	0.75	0.102	2.10	6.90	CAUTION: SMOKE IS EXITING OUT VENT
50	0.74	0.102	2.10	6.90	CAUTION: SMOKE IS EXITING OUT VENT
55	0.74	0.103	2.10	6.90	CAUTION: SMOKE IS EXITING OUT VENT
60	0.73	0.104	2.10	6.90	CAUTION: SMOKE IS EXITING OUT VENT

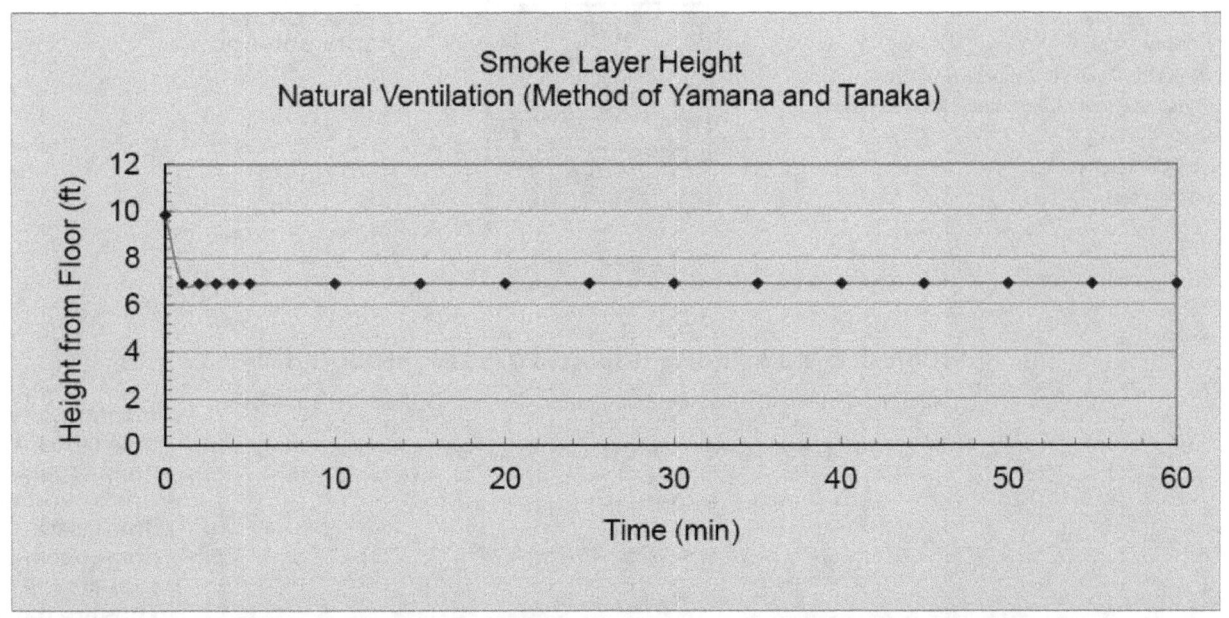

NOTE:
The above calculations are based on principles developed in the SFPE Handbook of Fire Protection Engineering, 3rd Edition, 2002. Calculations are based on certain assumptions and have inherent limitations. The results of such calculations may or may not have reasonable predictive capabilities for a given situation and should only be interpreted by an informed user. Although each calculation in the spreadsheet has been verified with the results of hand calculation, there is no absolute guarantee of the accuracy of these calculations. Any questions, comments, concerns, and suggestions, or to report an error(s) in the spreadsheet, please send an email to David.Stroup@nrc.gov and Naeem.Iqbal@nrc.gov or MarkHenry.Salley@nrc.gov.

The following calculations estimate the time to failure of cables exposed to a specified hot gas layer.

Parameters in YELLOW CELLS are Entered by the User.

Parameters in GREEN CELLS are Automatically Selected from the DROP DOWN MENU or SELECT CABLE BUTTON.

All subsequent output values are calculated by the spreadsheet and based on values specified in the input parameters. This spreadsheet is protected and secure to avoid errors due to a wrong entry in a cell(s). The chapter in the NUREG should be read before an analysis is made.

Project / Inspection Title:	NUREG-1805 Supplement 1 Example 19.11-3a

INPUT PARAMETERS

Cable Diameter	0.59	in
Cable Mass per Unit Length	0.26	b/ft
Cable Jacket Thickness	0.045	in
Ambient Air Temperature	70	°F
Failure Temperature	392	°F
Maximum Time	4000	s
Conduit Thickness	0.00	in
Conduit Outside Diameter	0.00	in
Cable Density	2151.45	kg/m³
Cable Insulation Type (Thermoplastic or Thermoset)	Thermoplastic	
Cable Function (Control, Instrumentation or Power)	Control	
Wire Gauge (AWG)	12	
Number of Conductors	7	
Cable Location	Cable Tray	

Do Not Enter Any Values in the Green Boxes!

They are entered automatically based on the cable selection.

Select Cable

Click to Select Source for Exposure Gas Temperature Profile:

Natural Ventilation - Method of McCaffrey, Quintiere, Harkleroad (MQH)
Forced Ventilation - Me hod of Foote, Pagni, and Alvares (FPA)
Forced Ven ilation - Me hod of Deal and Beyler
Room Fire with Closed Door
Within Fire Plume
User Defined

Selecting one of these items will automatically transfer you to the appropriate spreadsheet to calculate the exposure gas temperature profile

Warning: You MUST Click the Calculate Button Below when Finished Entering or Changing Data!

Calculate

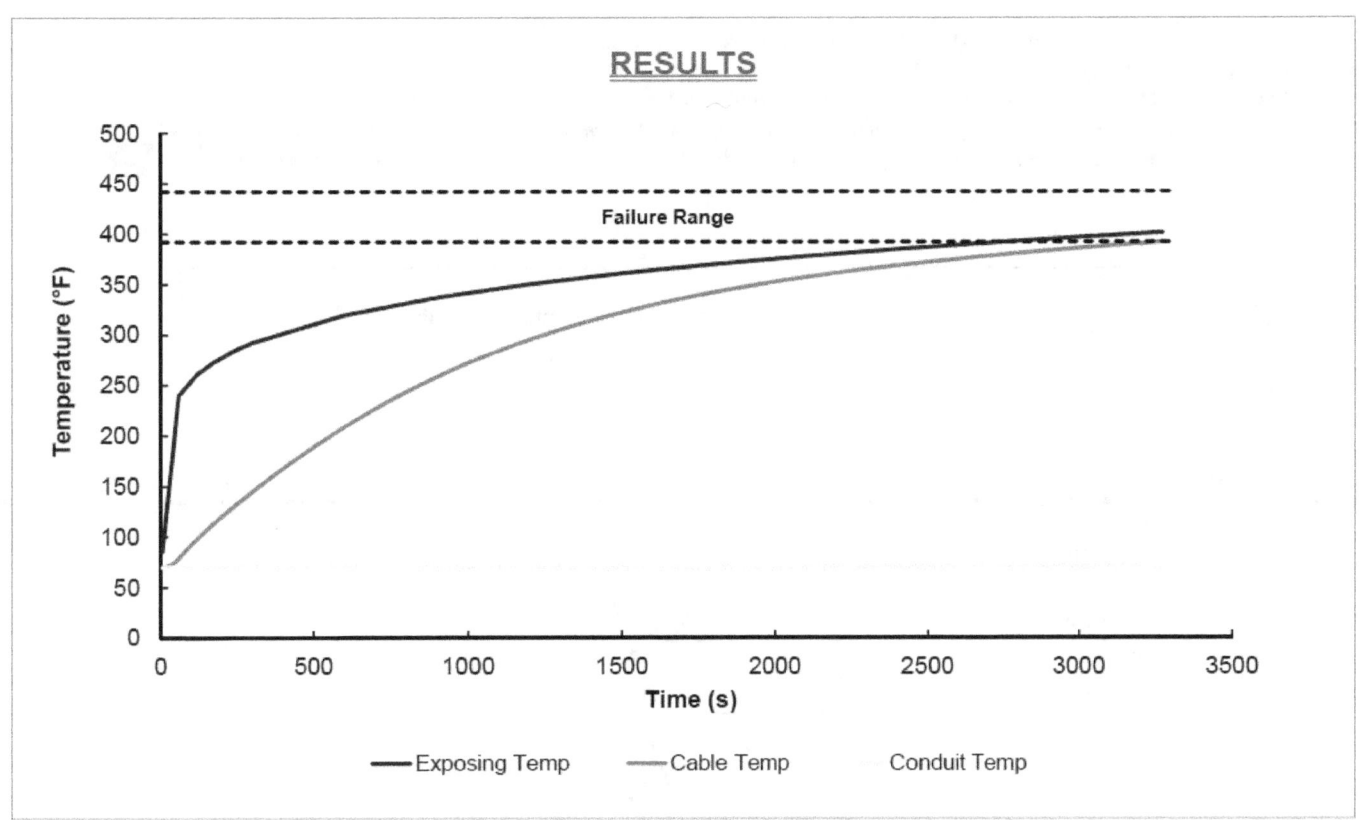

RESULTS

EXPOSURE GAS TEMPERATURE PROFILE		
Time (s)	Gas Temperature (°F)	Gas Temperature (°K)
0	70.00	294.26
60	240.10	388.76
120	260.93	400.33
180	274.28	407.75
240	284.31	413.32
300	292.43	417.83
600	319.67	432.97
900	337.13	442.67
1200	350.25	449.95
1500	360.87	455.85
1800	369.84	460.84
2100	377.64	465.17
2400	384.57	469.02
2700	390.80	472.48
3000	396.49	475.64
3300	401.71	478.55
3600	406.56	481.24

Answer: Cable fails at 54.6 minutes

COMPARTMENT WITH THERMALLY THICK/THIN BOUNDARIES

The following calculations estimate the hot gas layer temperature and smoke layer height in enclosure fire.

Parameters in YELLOW CELLS are Entered by the User.

Parameters in GREEN CELLS are Automatically Selected from the DROP DOWN MENU for the Material Selected.

All subsequent output values are calculated by the spreadsheet and based on values specified in the input parameters. This spreadsheet is protected and secure to avoid errors due to a wrong entry in a cell(s). The chapter in the NUREG should be read before an analysis is made.

**Project / Inspection
Title:**

NUREG-1805 Supplement 1 Example 19-11.3b

INPUT PARAMETERS

COMPARTMENT INFORMATION

Compartment Width (w_c)	16.40	ft
Compartment Length (l_c)	32.81	ft
Compartment Height (h_c)	9.84	ft
Vent Width (w_v)	3.28	ft
Vent Height (h_v)	6.90	ft
Top of Vent from Floor (V_T)	6.90	ft
Interior Lining Thickness (δ)	12.00	in

AMBIENT CONDITIONS

Ambient Air Temperature (T_a)	70.00	°F
Specific Heat of Air (c_p)	1.00	kJ/kg-K
Ambient Air Density (ρ_a)	1.20	kg/m³

Note: Ambient Air Density (ρ_a) will automatically correct with Ambient Air Temperature (T_a) Input

THERMAL PROPERTIES OF COMPARTMENT ENCLOSING SURFACES FOR

Interior Lining Thermal Inertia ($k\rho c$)	2.9	(kW/m²-K)²-sec
Interior Lining Thermal Conductivity (k)	0.0016	kW/m-K
Interior Lining Specific Heat (c)	0.75	kJ/kg-K
Interior Lining Density (ρ)	2400	kg/m³

CHAPTER 2. PREDICTING HOT GAS LAYER TEMPERATURE
AND SMOKE LAYER HEIGHT IN A ROOM FIRE
WITH NATURAL VENTILATION

Version 1805.1
(English Units)

THERMAL PROPERTIES FOR COMMON INTERIOR LINING MATERIALS

Material	kρc $(kW/m^2\text{-}K)^2\text{-sec}$	k (kW/m-K)	c (kJ/kg-K)	ρ (kg/m^3)	Select Material
					Concrete
Aluminum (pure)	500	0.206	0.895	2710	Scroll to desired material
Steel (0.5% Carbon)	197	0.054	0.465	7850	Click the selection
Concrete	2.9	0.0016	0.75	2400	
Brick	1.7	0.0008	0.8	2600	
Glass, Plate	1.6	0.00076	0.8	2710	
Brick/Concrete Block	1.2	0.00073	0.84	1900	
Gypsum Board	0.18	0.00017	1.1	960	
Plywood	0.16	0.00012	2.5	540	
Fiber Insulation Board	0.16	0.00053	1.25	240	
Chipboard	0.15	0.00015	1.25	800	
Aerated Concrete	0.12	0.00026	0.96	500	
Plasterboard	0.12	0.00016	0.84	950	
Calcium Silicate Board	0.098	0.00013	1.12	700	
Alumina Silicate Block	0.036	0.00014	1	260	
Glass Fiber Insulation	0.0018	0.000037	0.8	60	
Expanded Polystyrene	0.001	0.000034	1.5	20	
User Specified Value	Enter Value	Enter Value	Enter Value	Enter Value	

Reference: Klote, J., J. Milke, Principles of Smoke Management, 2002, Page 270.

FIRE SPECIFICATIONS

Fire Heat Release Rate (Q)   575.00 kW

**Click Here to Calculate this Sheet
and Return to THIEF**

Results	Time After Ignition (t)		h_k	ΔT_g	T_g	T_g	T_g
	(min)	(sec)	(kW/m²-K)	(°K)	(°K)	(°C)	(°F)
	0	0.00	-	-	294.11	21.11	70.00
	1	60	0.22	94.50	388.61	115.61	240.10
	2	120	0.16	106.07	400.18	127.18	260.93
	3	180	0.13	113.49	407.60	134.60	274.28
	4	240	0.11	119.06	413.17	140.17	284.31
	5	300	0.10	123.57	417.68	144.68	292.43
	10	600	0.07	138.71	432.82	159.82	319.67
	15	900	0.06	148.40	442.52	169.52	337.13
	20	1200	0.05	155.69	449.80	176.80	350.25
	25	1500	0.04	161.59	455.70	182.70	360.87
	30	1800	0.04	166.58	460.69	187.69	369.84
	35	2100	0.04	170.91	465.02	192.02	377.64
	40	2400	0.03	174.76	468.87	195.87	384.57
	45	2700	0.03	178.22	472.33	199.33	390.80
	50	3000	0.03	181.38	475.49	202.49	396.49
	55	3300	0.03	184.29	478.40	205.40	401.71
	60	3600	0.03	186.98	481.09	208.09	406.56

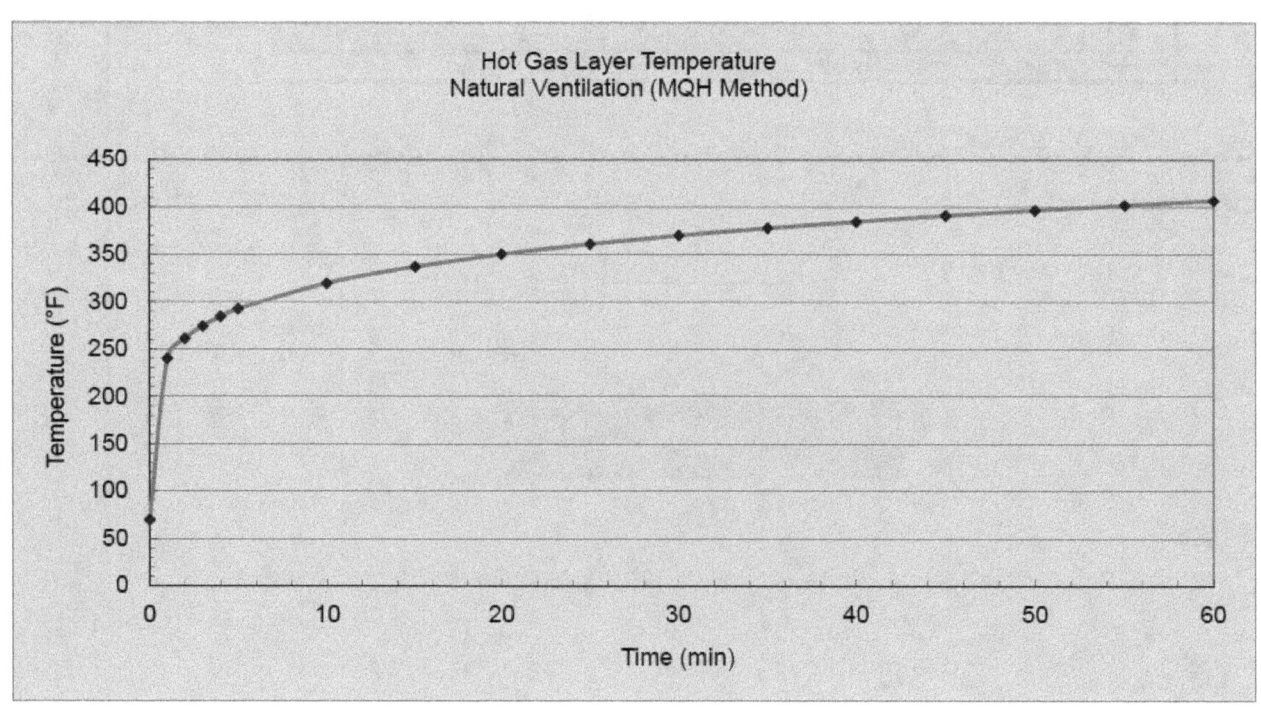

Results Caution! The smoke layer height is a conservative estimate and is only intended to provide an indication where the hot gas layer is located. Calculated smoke layer height below the vent height are not creditable since the calculation is not accounting for the smoke exiting the vent.

Time (min)	ρ_g (kg/m³)	Constant (k) (kW/m-K)	Smoke Layer Height z (m)	Smoke Layer Height z (ft)	
0	1.20	0.063	3.00	9.84	
1	0.91	0.084	2.10	6.90	CAUTION: SMOKE IS EXITING OUT VENT
2	0.88	0.086	2.10	6.90	CAUTION: SMOKE IS EXITING OUT VENT
3	0.87	0.088	2.10	6.90	CAUTION: SMOKE IS EXITING OUT VENT
4	0.85	0.089	2.10	6.90	CAUTION: SMOKE IS EXITING OUT VENT
5	0.85	0.090	2.10	6.90	CAUTION: SMOKE IS EXITING OUT VENT
10	0.82	0.093	2.10	6.90	CAUTION: SMOKE IS EXITING OUT VENT
15	0.80	0.095	2.10	6.90	CAUTION: SMOKE IS EXITING OUT VENT
20	0.78	0.097	2.10	6.90	CAUTION: SMOKE IS EXITING OUT VENT
25	0.77	0.098	2.10	6.90	CAUTION: SMOKE IS EXITING OUT VENT
30	0.77	0.099	2.10	6.90	CAUTION: SMOKE IS EXITING OUT VENT
35	0.76	0.100	2.10	6.90	CAUTION: SMOKE IS EXITING OUT VENT
40	0.75	0.101	2.10	6.90	CAUTION: SMOKE IS EXITING OUT VENT
45	0.75	0.102	2.10	6.90	CAUTION: SMOKE IS EXITING OUT VENT
50	0.74	0.102	2.10	6.90	CAUTION: SMOKE IS EXITING OUT VENT
55	0.74	0.103	2.10	6.90	CAUTION: SMOKE IS EXITING OUT VENT
60	0.73	0.104	2.10	6.90	CAUTION: SMOKE IS EXITING OUT VENT

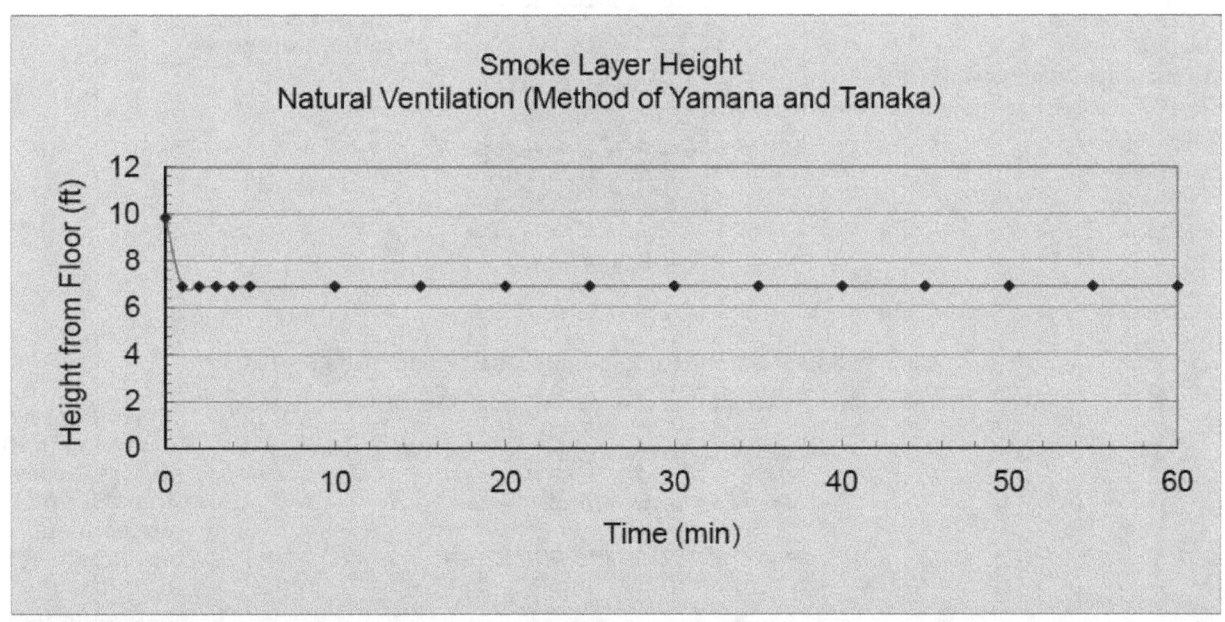

Smoke Layer Height
Natural Ventilation (Method of Yamana and Tanaka)

NOTE:
The above calculations are based on principles developed in the SFPE Handbook of Fire Protection Engineering, 3rd Edition, 2002. Calculations are based on certain assumptions and have inherent limitations. The results of such calculations may or may not have reasonable predictive capabilities for a given situation and should only be interpreted by an informed user. Although each calculation in the spreadsheet has been verified with the results of hand calculation, there is no absolute guarantee of the accuracy of these calculations. Any questions, comments, concerns, and suggestions, or to report an error(s) in the spreadsheet, please send an email to David.Stroup@nrc.gov and Naeem.Iqbal@nrc.gov or MarkHenry.Salley@nrc.gov.

The following calculations estimate the time to failure of cables exposed to a specified hot gas layer.

Parameters in YELLOW CELLS are Entered by the User.

Parameters in GREEN CELLS are Automatically Selected from the DROP DOWN MENU or SELECT CABLE BUTTON.

All subsequent output values are calculated by the spreadsheet and based on values specified in the input parameters. This spreadsheet is protected and secure to avoid errors due to a wrong entry in a cell(s). The chapter in the NUREG should be read before an analysis is made.

Project / Inspection Title:	NUREG-1805 Supplement 1 Example 19.11-3b

INPUT PARAMETERS

Cable Diameter	0.59	in
Cable Mass per Unit Length	0.26	b/ft
Cable Jacket Thickness	0.045	in
Ambient Air Temperature	70	°F
Failure Temperature	392	°F
Maximum Time	4000	s
Conduit Thickness	0.21	in
Conduit Outside Diameter	3.50	in
Cable Density	2151.45	kg/m³
Cable Insulation Type (Thermoplastic or Thermoset)	Thermoplastic	
Cable Function (Control, Instrumentation or Power)	Control	
Wire Gauge (AWG)	12	
Number of Conductors	7	
Cable Location	Conduit - Rigid	

Do Not Enter Any Values in the Green Boxes!

They are entered automatically based on the cable selection.

Select Cable

Click to Select Source for Exposure Gas Temperature Profile:

Natural Ventilation - Method of McCaffrey, Quintiere, Harkleroad (MQH)
Forced Ventilation - Method of Foote, Pagni, and Alvares (FPA)
Forced Ventilation - Method of Deal and Beyler
Room Fire with Closed Door
Within Fire Plume
User Defined

Selecting one of these items will automatically transfer you to the appropriate spreadsheet to calculate the exposure gas temperature profile

Warning: You MUST Click the Calculate Button Below when Finished Entering or Changing Data!

Calculate

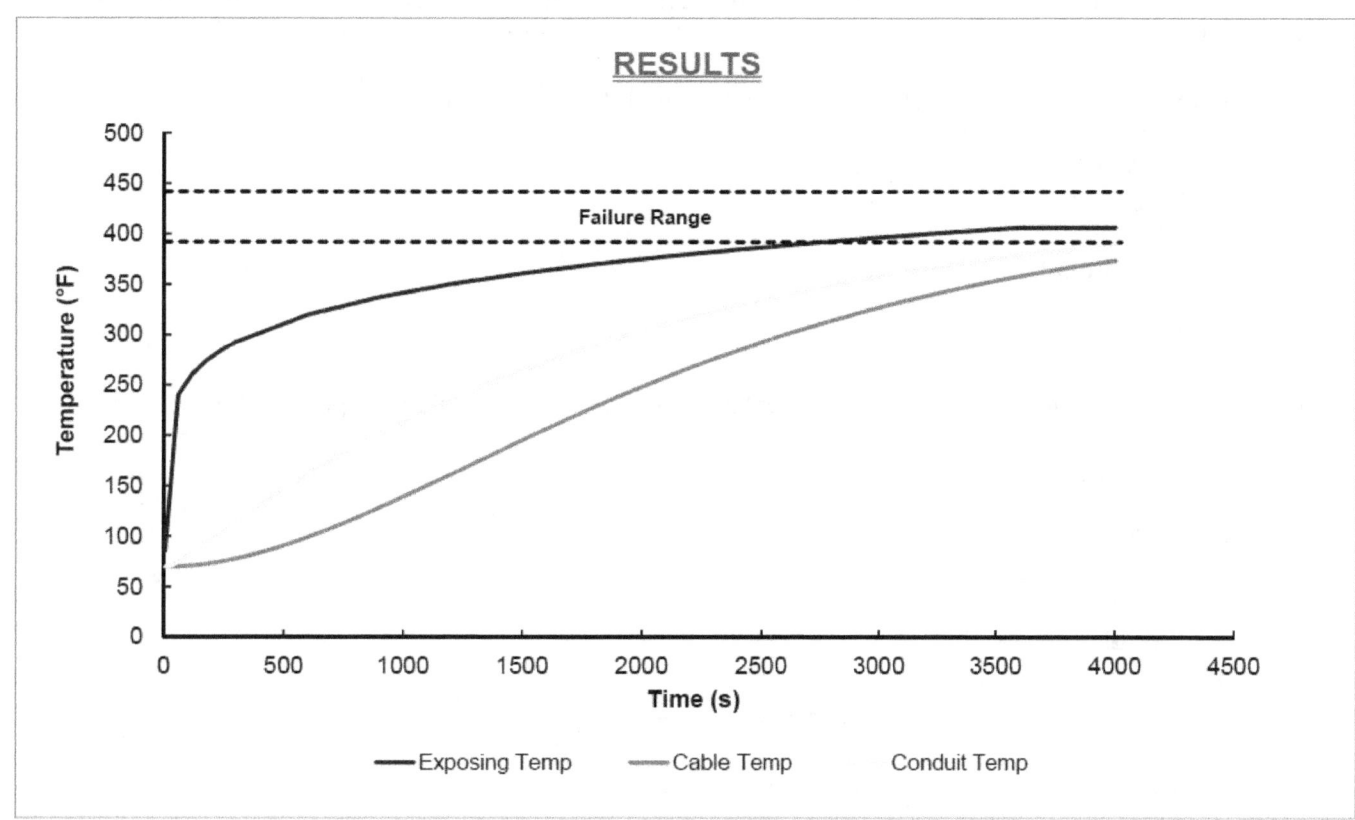

EXPOSURE GAS TEMPERATURE PROFILE		
Time (s)	Gas Temperature (°F)	Gas Temperature (°K)
0	70.00	294.26
60	240.10	388.76
120	260.93	400.33
180	274.28	407.75
240	284.31	413.32
300	292.43	417.83
600	319.67	432.97
900	337.13	442.67
1200	350.25	449.95
1500	360.87	455.85
1800	369.84	460.84
2100	377.64	465.17
2400	384.57	469.02
2700	390.80	472.48
3000	396.49	475.64
3300	401.71	478.55
3600	406.56	481.24

Answer: Cable does not reach failure temperature in 4000.3 seconds

Example Problem 19.11-4
(Cable Type – Thermoplastic vs. Thermoset – Forced Ventilation)

Problem Statement
Consider a concrete compartment that is 16.40 ft (5.00 m) wide x 16.40 ft (5.00 m) long x 11.48 ft (3.50 m) high (w_c x l_c x h_c) and forced ventilation at a rate of 1000.00 cfm. The fire is constant with a HRR of 500kW. Compute the time to cable failure under two scenarios:

(a)   Thermoplastic cable located in a cable tray with the following parameters
  -   Control
  -   14 AWG
  -   9 conductors
  -   PE insulation
  -   PVC jacket
  -   Dekoron
  -   Model : 1735

(b)   Thermoset cable is located in the same cable tray with the following parameters
  -   Control
  -   14 AWG
  -   9 conductors
  -   XLPE insulation
  -   CSPE jacket
  -   Rockbestos-Surprenant
  -   Model : Firewall III [®]

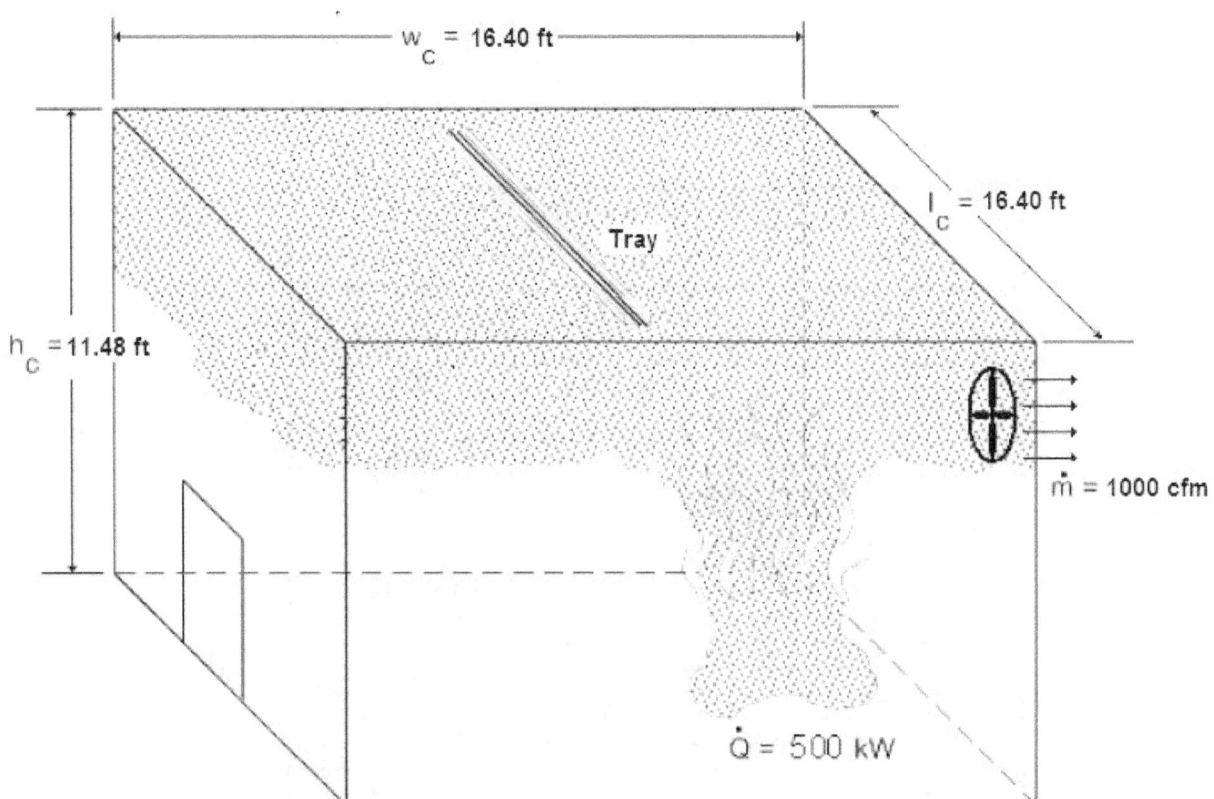

Example 19.11-4. Concrete Compartment with Forced Ventilation.

Solution

Purpose:

    (1)   Determine the time to failure and compare the results of cables with similar physical construction, but different insulation (differing chemical properties).

Assumptions:

    (1)   The cable is surrounded by a uniform temperature hot gas layer.
    (2)   The cable properties do not change with temperature increase.
    (3)   Failure is indicated when the temperature inside the cable jacket exceeds 392°F (200°C) for the thermoplastic cable and 752°F (400°C) for the thermoset cable.

Spreadsheet (FDT^s) Information:

Use the following FDT^s:
    (a)   02.2_Temperature_FV_Sup1.xls (click on *Temperature_FV*)
    (b)   19_THIEF_Thermally_Induced_Electrical_Failure_of_Cables_ Sup1.xls (click on *THIEF*)

FDT^s Input Parameters
    -Gas temperature around the cable as a function time:
    Use 02.2_Temperature_FV_Sup1.xls and the parameters presented above to determine the gas temperature (same for both scenarios)
    On the THIEF spreadsheet, select: Forced Ventilation – Method of Deal and Beyler or Forced Ventilation – Method of Foote, Pagni, and Alvares (FPA)
    -Press "Select Cable" Button
    -Choose appropriate cable from list
        (a) Control – 14 AWG – 9 conductor – Dekoron – 1735 – PE – PVC
        (b) Control – 14 AWG – 9 conductor – Rockbestos-Surprenant – XLPE - CSPE
    -Choose appropriate Cable Location ("Cable Tray")
    -Press "Calculate" Button

Results*

	FPA Method	Deal and Beyler
Thermoplastic cable failure time (minutes)	31.8	30.9
Thermoset cable failure time (minutes)	No Failure	No Failure

*see spreadsheets

The results show a thermoplastic cable failing prior to a thermoset cable of the same cable conductor configuration. This example illustrates the added thermal protection of a thermoset (TS) cable relative to a less thermally robust thermoplastic (TP) cable. The difference is a direct result of the difference between formulations of the polymeric insulating materials used in the two different cables.

COMPARTMENT WITH THERMALLY THICK/THIN BOUNDARIES

The following calculations estimate the hot gas layer temperature and smoke layer height in enclosure fire.

Parameters in YELLOW CELLS are Entered by the User.

Parameters in GREEN CELLS are Automatically Selected from the DROP DOWN MENU for the Material Selected.

All subsequent output values are calculated by the spreadsheet and based on values specified in the input parameters. This spreadsheet is protected and secure to avoid errors due to a wrong entry in a cell(s). The chapter in the NUREG should be read before an analysis is made.

Project / Inspection Title:	NUREG-1805 Supplement 1 Example 19-11.4a

INPUT PARAMETERS

COMPARTMENT INFORMATION

Compartment Width (w_c)	16.40	ft
Compartment Length (l_c)	16.40	ft
Compartment Height (h_c)	11.48	ft
Interior Lining Thickness (δ)	12.00	in

AMBIENT CONDITIONS

Ambient Air Temperature (T_a)	70.00	°F
Specific Heat of Air (c_p)	1.00	kJ/kg-K
Ambient Air Density (ρ_a)	1.20	kg/m^3

Note: Ambient Air Density (ρ_a) will automatically correct with Ambient Air Temperature (T_a) Input

THERMAL PROPERTIES OF COMPARTMENT ENCLOSING SURFACES FOR

Interior Lining Thermal Inertia ($k\rho c$)	2.9	(kW/m^2-K)^2-sec
Interior Lining Thermal Conductivity (k)	0.0016	kW/m-K
Interior Lining Specific Heat (c)	0.75	kJ/kg-K
Interior Lining Density (ρ)	2400	kg/m^3

THERMAL PROPERTIES FOR COMMON INTERIOR LINING MATERIALS

Material	kρc (kW/m²-K)²-sec	k (kW/m-K)	c (kJ/kg-K)	ρ (kg/m³)	Select Material
					Concrete ▼
Aluminum (pure)	500	0.206	0.895	2710	Scroll to desired material
Steel (0.5% Carbon)	197	0.054	0.465	7850	Click on selection
Concrete	2.9	0.0016	0.75	2400	
Brick	1.7	0.0008	0.8	2600	
Glass, Plate	1.6	0.00076	0.8	2710	
Brick/Concrete Block	1.2	0.00073	0.84	1900	
Gypsum Board	0.18	0.00017	1.1	960	
Plywood	0.16	0.00012	2.5	540	
F ber Insulation Board	0.16	0.00053	1.25	240	
Chipboard	0.15	0.00015	1.25	800	
Aerated Concrete	0.12	0.00026	0.96	500	
Plasterboard	0.12	0.00016	0.84	950	
Calcium Silicate Board	0.098	0.00013	1.12	700	
Alumina Silicate Block	0.036	0.00014	1	260	
Glass Fiber Insulation	0.0018	0.000037	0.8	60	
Expanded Polystyrene	0.001	0.000034	1.5	20	
User Specified Value	Enter Value	Enter Value	Enter Value	Enter Value	

Reference: Klote, J., J. Milke, Principles of Smoke Management, 2002 Page 270.

COMPARTMENT MASS VENTILATION FLOW RATE

Forced Ventilation Flow Rate (m)   1000.00 cfm

FIRE SPECIFICATIONS

Fire Heat Release Rate (Q)   500.00 kW

Click Here to Calculate this Sheet
and Return to THIEF

Compartment Hot Gas Layer Temperature With Forced Ventilation

$$\Delta T_g/T_a = 0.63(Q/mc_pT_a)^{0.72}(h_kA_T/mc_p)^{-0.36}$$

$$\Delta T_g = T_g - T_a$$

$$T_g = \Delta T_g + T_a$$

Results	Time After Ignition (t)		h_k	$\Delta T_g/T_a$	ΔT_g	T_g	T_g	T_g
	(min)	(sec)	(kW/m²-K)		(°K)	(°K)	(°C)	(°F)
	0	0	-	-	-	294.11	21.11	70.00
	1	60	0.22	0.35	102.58	396.69	123.69	254.64
	2	120	0.16	0.40	116.21	410.32	137.32	279.18
	3	180	0.13	0.43	125.01	419.12	146.12	295.01
	4	240	0.11	0.45	131.65	425.76	152.76	306.97
	5	300	0.10	0.47	137.05	431.16	158.16	316.69
	10	600	0.07	0.53	155.26	449.37	176.37	349.47
	15	900	0.06	0.57	167.01	461.12	188.12	370.62
	20	1200	0.05	0.60	175.89	470.00	197.00	386.60
	25	1500	0.04	0.62	183.10	477.21	204.21	399.58
	30	1800	0.04	0.64	189.21	483.32	210.32	410.57
	35	2100	0.04	0.66	194.53	488.64	215.64	420.16
	40	2400	0.03	0.68	199.26	493.37	220.37	428.67
	45	2700	0.03	0.69	203.53	497.64	224.64	436.36
	50	3000	0.03	0.71	207.43	501.54	228.54	443.37
	55	3300	0.03	0.72	211.02	505.13	232.13	449.83
	60	3600	0.03	0.73	214.35	508.46	235.46	455.83

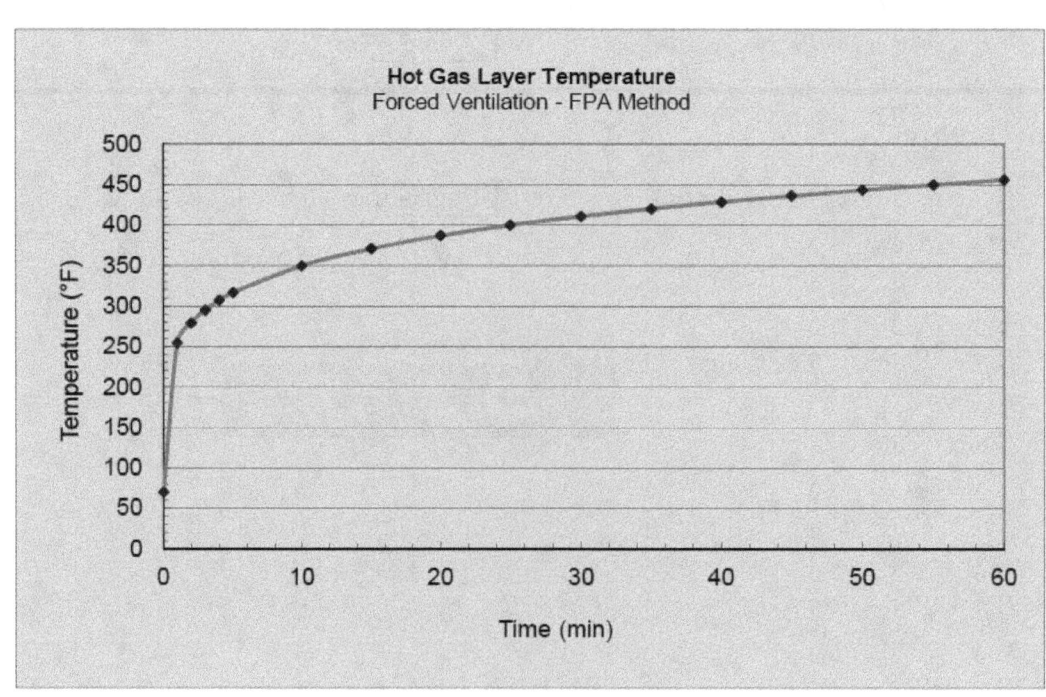

Hot Gas Layer Temperature
Forced Ventilation - FPA Method

Time After Ignition (t)		h_k	ΔT_g	T_g	T_g	T_g
(min)	(sec)	(kW/m²-K)	(°K)	(°K)	(°C)	(°F)
0	0	-	-	294.11	21.11	70.00
1	60	0.09	44.99	339.10	66.10	150.98
2	120	0.06	62.31	356.42	83.42	182.16
3	180	0.05	75.12	369.23	96.23	205.22
4	240	0.04	85.61	379.73	106.73	224.11
5	300	0.04	94.64	388.75	115.75	240.35
10	600	0.03	128.15	422.26	149.26	300.66
15	900	0.02	151.99	446.10	173.10	343.58
20	1200	0.02	170.95	465.06	192.06	377.70
25	1500	0.02	186.85	480.96	207.96	406.33
30	1800	0.02	200.63	494.74	221.74	431.14
35	2100	0.01	212.83	506.94	233.94	453.10
40	2400	0.01	223.80	517.91	244.91	472.84
45	2700	0.01	233.78	527.89	254.89	490.81
50	3000	0.01	242.95	537.06	264.06	507.30
55	3300	0.01	251.43	545.54	272.54	522.57
60	3600	0.01	259.32	553.43	280.43	536.78

Results

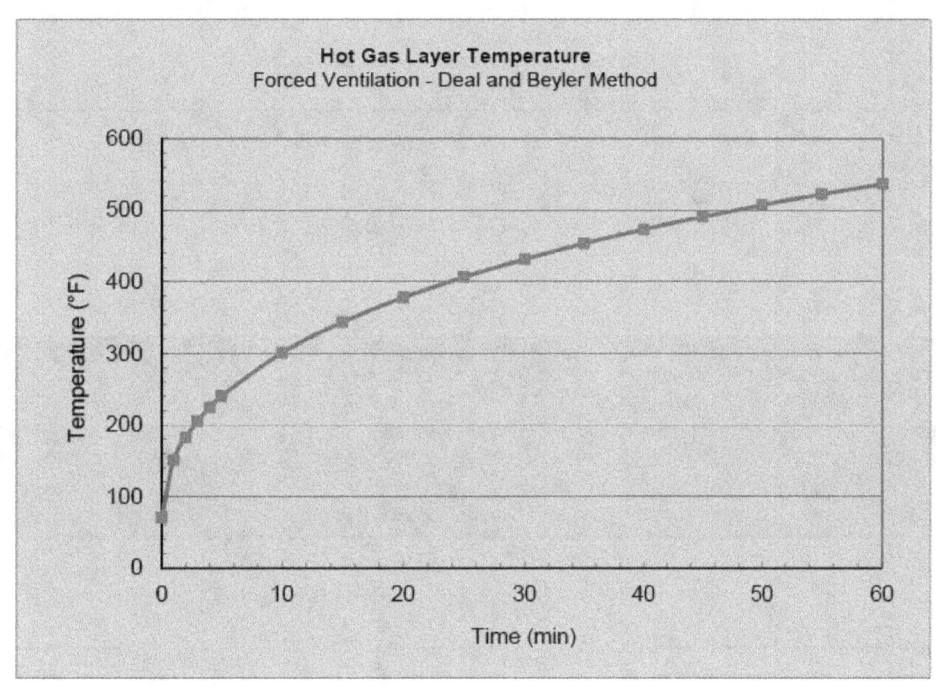

Summary of Results

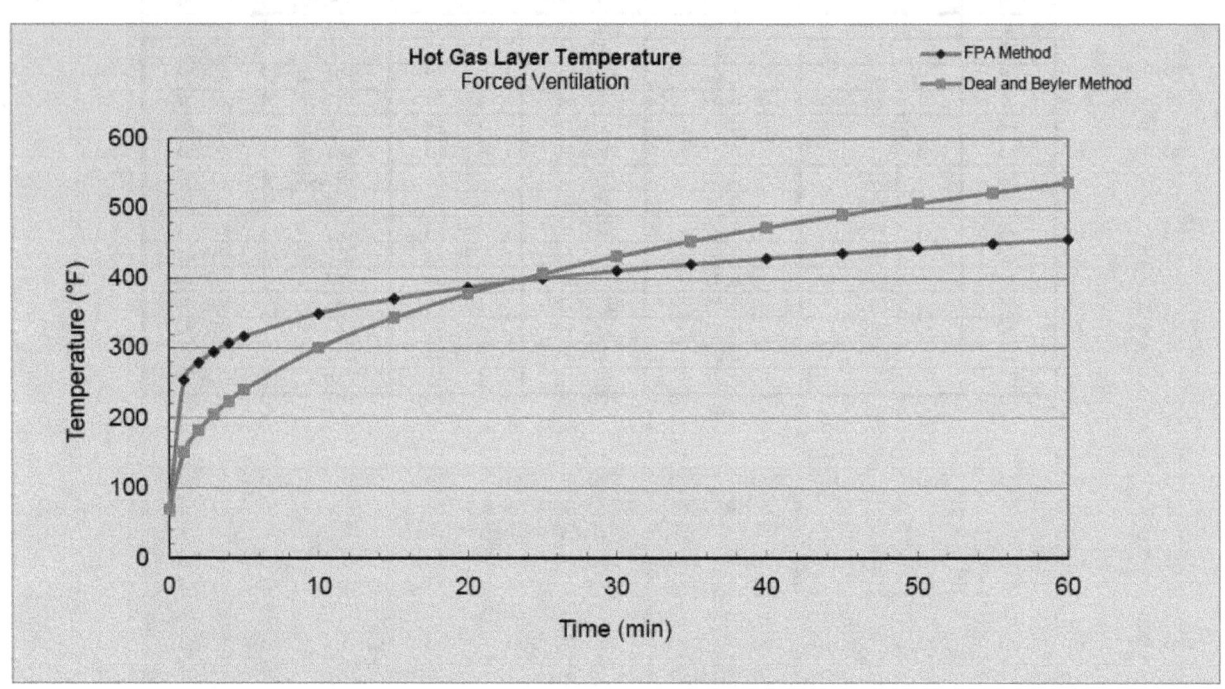

NOTE:
The above calculations are based on principles developed in the SFPE Handbook of Fire Protection Engineering, 2nd Edition, 1995. Calculations are based on certain assumptions and have inherent limitations. The results of such calculations may or may not have reasonable predictive capabilities for a given situation, and should only be interpreted by an informed user. Although each calculation in the spreadsheet has been verified with the results of hand calculation, there is no absolute guarantee of the accuracy of these calculations. Any questions, comments, concerns, and suggestions, or to report an error(s) in the spreadsheets, please send an email to David.Stroup@nrc.gov and Naeem.Iqbal@nrc.gov or MarkHenry.Salley@nrc.gov.

Prepared by: [   ]   Date: [   ]   Organization: [   ]

Checked by: [   ]   Date: [   ]   Organization: [   ]

Additional Information:

The following calculations estimate the time to failure of cables exposed to a specified hot gas layer.

Parameters in YELLOW CELLS are Entered by the User.

Parameters in GREEN CELLS are Automatically Selected from the DROP DOWN MENU or SELECT CABLE BUTTON.

All subsequent output values are calculated by the spreadsheet and based on values specified in the input parameters. This spreadsheet is protected and secure to avoid errors due to a wrong entry in a cell(s). The chapter in the NUREG should be read before an analysis is made.

Project / Inspection Title:	NUREG-1805 Supplement 1 Example 19.11-4a

INPUT PARAMETERS

Cable Diameter	0.65	in
Cable Mass per Unit Length	0.25	b/ft
Cable Jacket Thickness	0.060	in
Ambient Air Temperature	70	°F
Failure Temperature	392	°F
Maximum Time	4000	s
Conduit Thickness	0.00	in
Conduit Outside Diameter	0.00	in
Cable Density	1723.93	kg/m³
Cable Insulation Type (Thermoplastic or Thermoset)	Thermoplastic	
Cable Function (Control, Instrumentation or Power)	Control	
Wire Gauge (AWG)	14	
Number of Conductors	9	
Cable Location	Cable Tray	

Do Not Enter Any Values in the Green Boxes!

They are entered automatically based on the cable selection.

Select Cable

Click to Select Source for Exposure Gas Temperature Profile:

Natural Ventilation - Method of McCaffrey, Quintiere, Harkleroad (MQH)
Forced Ventilation - Method of Foote, Pagni, and Alvares (FPA)
Forced Ventilation - Method of Deal and Beyler
Room Fire with Closed Door
Within Fire Plume
User Defined

Selecting one of these items will automatically transfer you to the appropriate spreadsheet to calculate the exposure gas temperature profile

Warning: You MUST Click the Calculate Button Below when Finished Entering or Changing Data!

Calculate

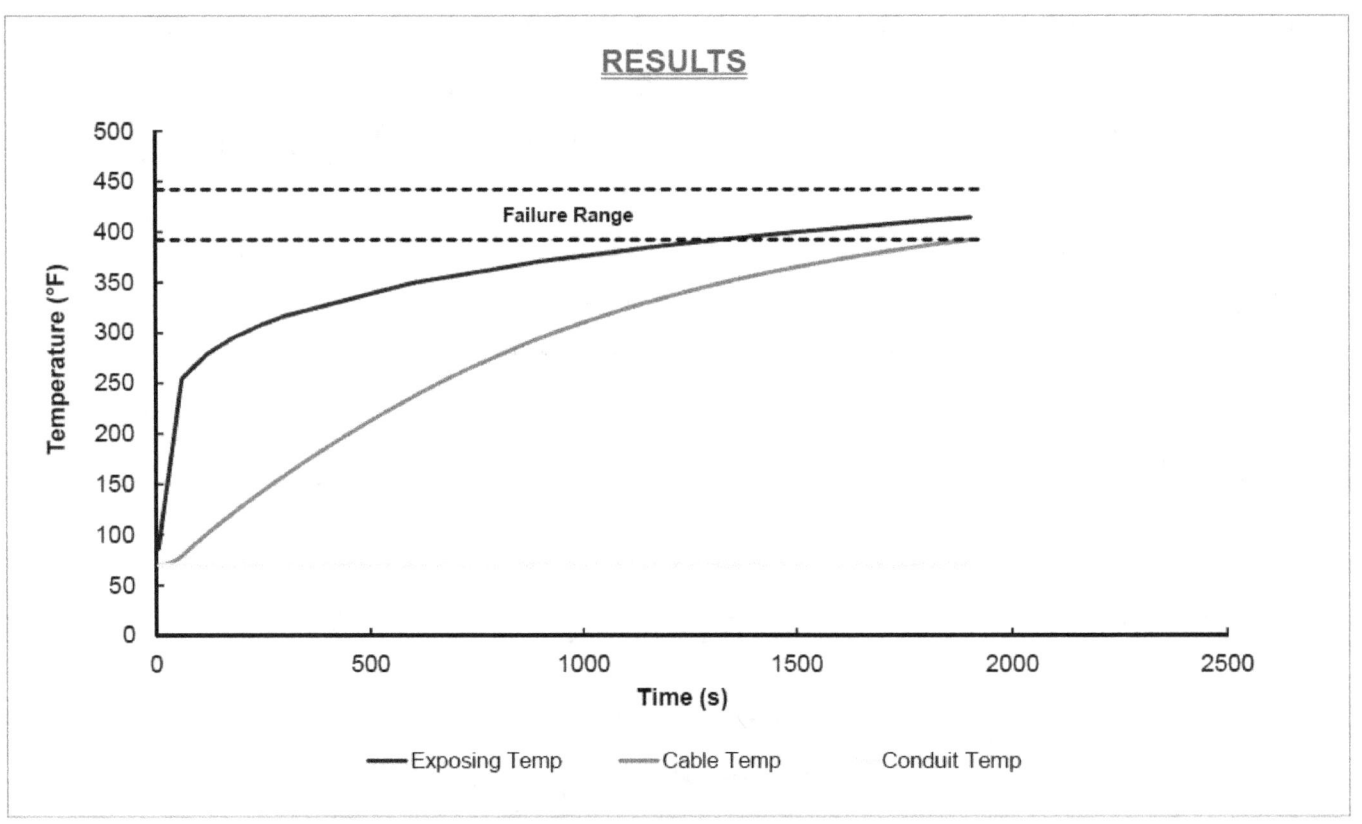

EXPOSURE GAS TEMPERATURE PROFILE		
Time (s)	Gas Temperature (°F)	Gas Temperature (°K)
0	70.00	294.26
60	254.64	396.84
120	279.18	410.47
180	295.01	419.27
240	306.97	425.91
300	316.69	431.31
600	349.47	449.52
900	370.62	461.27
1200	386.60	470.15
1500	399.58	477.36
1800	410.57	483.47
2100	420.16	488.79
2400	428.67	493.52
2700	436.36	497.79
3000	443.37	501.69
3300	449.83	505.28
3600	455.83	508.61

Answer: Cable fails at 31.8 minutes

COMPARTMENT WITH THERMALLY THICK/THIN BOUNDARIES

The following calculations estimate the hot gas layer temperature and smoke layer height in enclosure fire.

Parameters in YELLOW CELLS are Entered by the User.

Parameters in GREEN CELLS are Automatically Selected from the DROP DOWN MENU for the Material Selected.

All subsequent output values are calculated by the spreadsheet and based on values specified in the input parameters. This spreadsheet is protected and secure to avoid errors due to a wrong entry in a cell(s). The chapter in the NUREG should be read before an analysis is made.

Project / Inspection Title:	NUREG-1805 Supplement 1 Example 19-11.4b

INPUT PARAMETERS

COMPARTMENT INFORMATION

Compartment Width (w_c)	16.40	ft
Compartment Length (l_c)	16.40	ft
Compartment Height (h_c)	11.48	ft
Interior Lining Thickness (δ)	12.00	in

AMBIENT CONDITIONS

Ambient Air Temperature (T_a)	70.00	°F
Specific Heat of Air (c_p)	1.00	kJ/kg-K
Ambient Air Density (ρ_a)	1.20	kg/m^3

Note: Ambient Air Density (ρ_a) will automatically correct with Ambient Air Temperature (T_a) Input

THERMAL PROPERTIES OF COMPARTMENT ENCLOSING SURFACES FOR

Interior Lining Thermal Inertia ($k\rho c$)	2.9	(kW/m^2-K)^2-sec
Interior Lining Thermal Conductivity (k)	0.0016	kW/m-K
Interior Lining Specific Heat (c)	0.75	kJ/kg-K
Interior Lining Density (ρ)	2400	kg/m^3

THERMAL PROPERTIES FOR COMMON INTERIOR LINING MATERIALS

Material	kρc (kW/m²-K)²-sec	k (kW/m-K)	c (kJ/kg-K)	ρ (kg/m³)	Select Material
					Concrete ▼
Aluminum (pure)	500	0.206	0.895	2710	**Scroll to desired material**
Steel (0.5% Carbon)	197	0.054	0.465	7850	**Click on selection**
Concrete	2.9	0.0016	0.75	2400	
Brick	1.7	0.0008	0.8	2600	
Glass, Plate	1.6	0.00076	0.8	2710	
Brick/Concrete Block	1.2	0.00073	0.84	1900	
Gypsum Board	0.18	0.00017	1.1	960	
Plywood	0.16	0.00012	2.5	540	
F ber Insulation Board	0.16	0.00053	1.25	240	
Chipboard	0.15	0.00015	1.25	800	
Aerated Concrete	0.12	0.00026	0.96	500	
Plasterboard	0.12	0.00016	0.84	950	
Calcium Silicate Board	0.098	0.00013	1.12	700	
Alumina Silicate Block	0.036	0.00014	1	260	
Glass Fiber Insulation	0.0018	0.000037	0.8	60	
Expanded Polystyrene	0.001	0.000034	1.5	20	
User Specified Value	Enter Value	Enter Value	Enter Value	Enter Value	

Reference: Klote, J., J. Milke, Principles of Smoke Management, 2002 Page 270.

COMPARTMENT MASS VENTILATION FLOW RATE

Forced Ventilation Flow Rate (m)   1000.00 cfm

FIRE SPECIFICATIONS

Fire Heat Release Rate (Q)   500.00 kW

Click Here to Calculate this Sheet
and Return to THIEF

CHAPTER 2. PREDICTING HOT GAS LAYER TEMPERATURE
IN A ROOM FIRE
WITH FORCED VENTILATION

Version 1805.1
(English Units)

Compartment Hot Gas Layer Temperature With Forced Ventilation

$$\Delta T_g/T_a = 0.63(Q/mc_pT_a)^{0.72}(h_kA_T/mc_p)^{-0.36}$$

$$\Delta T_g = T_g - T_a$$

$$T_g = \Delta T_g + T_a$$

Results	Time After Ignition (t)		h_k	$\Delta T_g/T_a$	ΔT_g	T_g	T_g	T_g
	(min)	(sec)	(kW/m²-K)		(°K)	(°K)	(°C)	(°F)
	0	0	-	-	-	294.11	21.11	70.00
	1	60	0.22	0.35	102.58	396.69	123.69	254.64
	2	120	0.16	0.40	116.21	410.32	137.32	279.18
	3	180	0.13	0.43	125.01	419.12	146.12	295.01
	4	240	0.11	0.45	131.65	425.76	152.76	306.97
	5	300	0.10	0.47	137.05	431.16	158.16	316.69
	10	600	0.07	0.53	155.26	449.37	176.37	349.47
	15	900	0.06	0.57	167.01	461.12	188.12	370.62
	20	1200	0.05	0.60	175.89	470.00	197.00	386.60
	25	1500	0.04	0.62	183.10	477.21	204.21	399.58
	30	1800	0.04	0.64	189.21	483.32	210.32	410.57
	35	2100	0.04	0.66	194.53	488.64	215.64	420.16
	40	2400	0.03	0.68	199.26	493.37	220.37	428.67
	45	2700	0.03	0.69	203.53	497.64	224.64	436.36
	50	3000	0.03	0.71	207.43	501.54	228.54	443.37
	55	3300	0.03	0.72	211.02	505.13	232.13	449.83
	60	3600	0.03	0.73	214.35	508.46	235.46	455.83

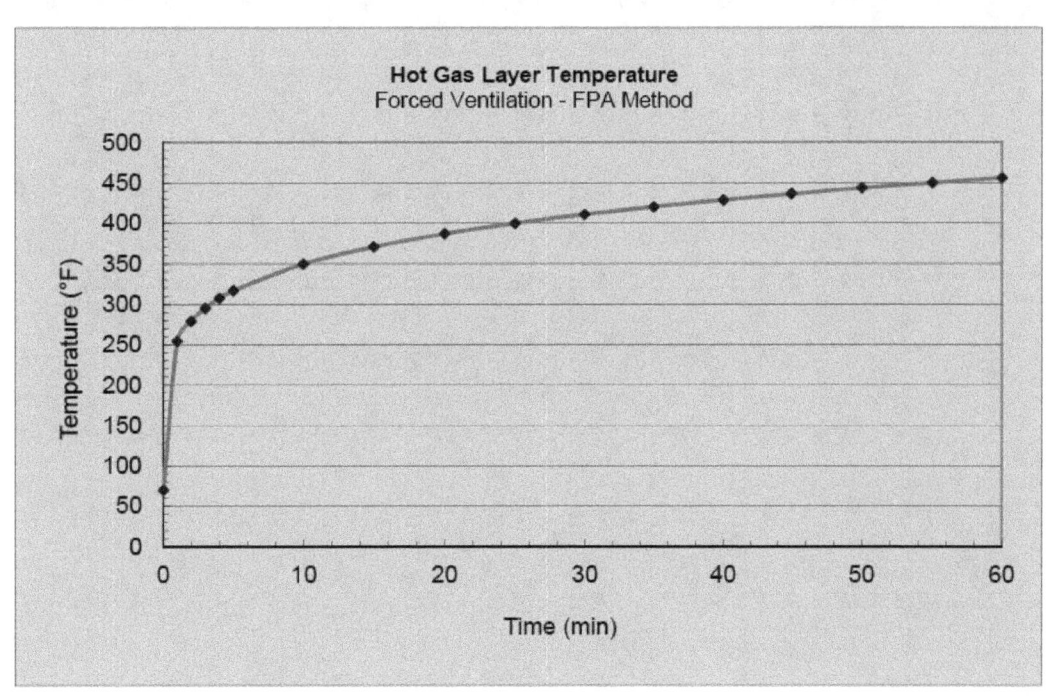

Hot Gas Layer Temperature
Forced Ventilation - FPA Method

	Time After Ignition (t)		h_k	ΔT_g	T_g	T_g	T_g
Results	(min)	(sec)	(kW/m²-K)	(°K)	(°K)	(°C)	(°F)
	0	0	-	-	294.11	21.11	70.00
	1	60	0.09	44.99	339.10	66.10	150.98
	2	120	0.06	62.31	356.42	83.42	182.16
	3	180	0.05	75.12	369.23	96.23	205.22
	4	240	0.04	85.61	379.73	106.73	224.11
	5	300	0.04	94.64	388.75	115.75	240.35
	10	600	0.03	128.15	422.26	149.26	300.66
	15	900	0.02	151.99	446.10	173.10	343.58
	20	1200	0.02	170.95	465.06	192.06	377.70
	25	1500	0.02	186.85	480.96	207.96	406.33
	30	1800	0.02	200.63	494.74	221.74	431.14
	35	2100	0.01	212.83	506.94	233.94	453.10
	40	2400	0.01	223.80	517.91	244.91	472.84
	45	2700	0.01	233.78	527.89	254.89	490.81
	50	3000	0.01	242.95	537.06	264.06	507.30
	55	3300	0.01	251.43	545.54	272.54	522.57
	60	3600	0.01	259.32	553.43	280.43	536.78

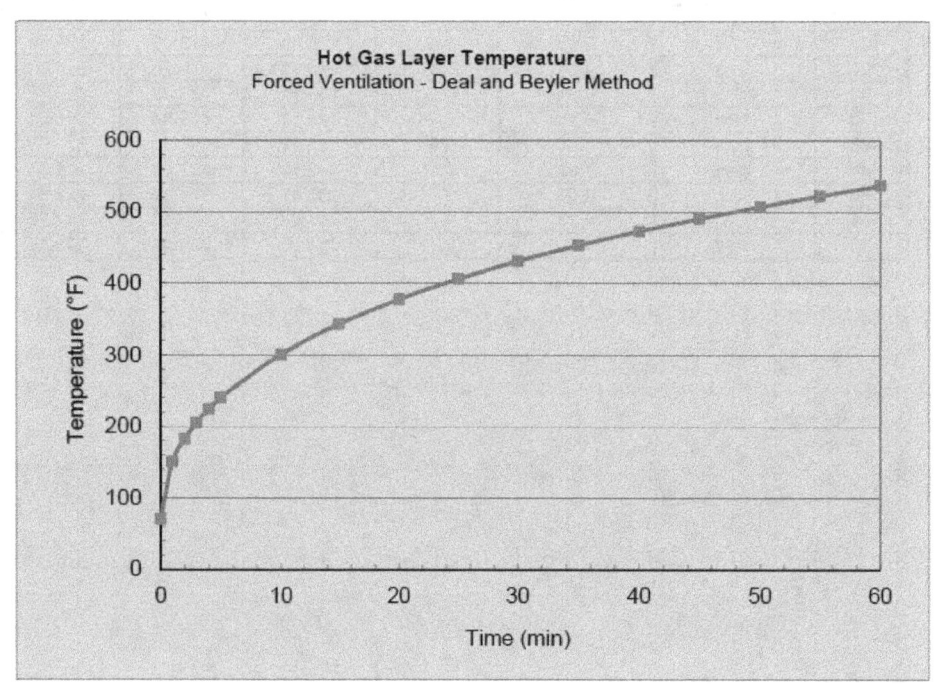

Summary of Results

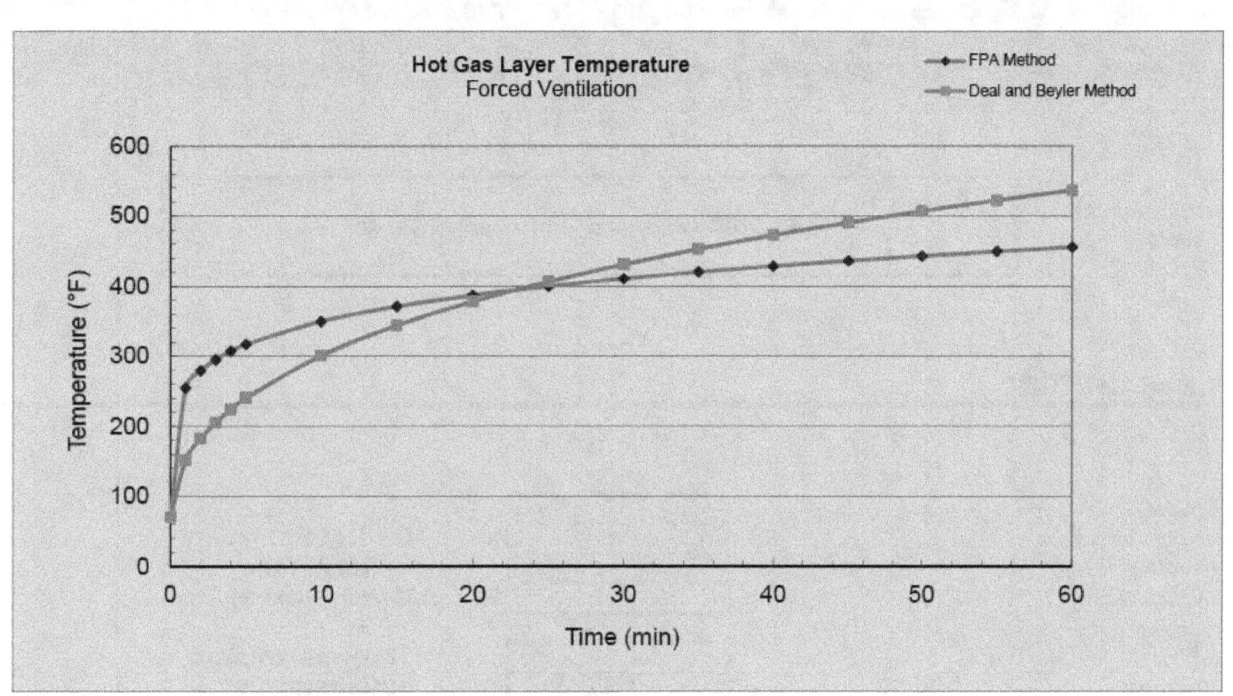

NOTE:
The above calculations are based on principles developed in the SFPE Handbook of Fire Protection Engineering, 2nd Edition, 1995. Calculations are based on certain assumptions and have inherent limitations. The results of such calculations may or may not have reasonable predictive capabilities for a given situation, and should only be interpreted by an informed user. Although each calculation in the spreadsheet has been verified with the results of hand calculation, there is no absolute guarantee of the accuracy of these calculations. Any questions, comments, concerns, and suggestions, or to report an error(s) in the spreadsheets, please send an email to David.Stroup@nrc.gov and Naeem.Iqbal@nrc.gov or MarkHenry.Salley@nrc.gov.

Prepared by: [   ]   Date: [   ]   Organization: [   ]

Checked by: [   ]   Date: [   ]   Organization: [   ]

Additional Information:

The following calculations estimate the time to failure of cables exposed to a specified hot gas layer.

Parameters in YELLOW CELLS are Entered by the User.

Parameters in GREEN CELLS are Automatically Selected from the DROP DOWN MENU or SELECT CABLE BUTTON.

All subsequent output values are calculated by the spreadsheet and based on values specified in the input parameters. This spreadsheet is protected and secure to avoid errors due to a wrong entry in a cell(s). The chapter in the NUREG should be read before an analysis is made.

Project / Inspection Title:	NUREG-1805 Supplement 1 Example 19.11-4b

INPUT PARAMETERS

Cable Diameter	0.65	in	**Do Not Enter Any**
Cable Mass per Unit Length	0.25	b/ft	**Values in the**
Cable Jacket Thickness	0.060	in	**Green Boxes!**
Ambient Air Temperature	70	°F	
Failure Temperature	392	°F	
Maximum Time	4000	s	**They are entered**
Conduit Thickness	0.00	in	**automatically**
Conduit Outside Diameter	0.00	in	**based on the**
Cable Density	1723.93	kg/m³	**cable selection.**
Cable Insulation Type (Thermoplastic or Thermoset)	Thermoplastic		
Cable Function (Control, Instrumentation or Power)	Control		
Wire Gauge (AWG)	14		
Number of Conductors	9		
Cable Location	Cable Tray		

Select Cable

Click to Select Source for Exposure Gas Temperature Profile:

Natural Ventilation - Method of McCaffrey, Quintiere, Harkleroad (MQH)
Forced Ventilation - Method of Foote, Pagni, and Alvares (FPA)
Forced Ventilation - Method of Deal and Beyler
Room Fire with Closed Door
Within Fire Plume
User Defined

Selecting one of these items will automatically transfer you to the appropriate spreadsheet to calculate the exposure gas temperature profile

Warning: You MUST Click the Calculate Button Below when Finished Entering or Changing Data!

Calculate

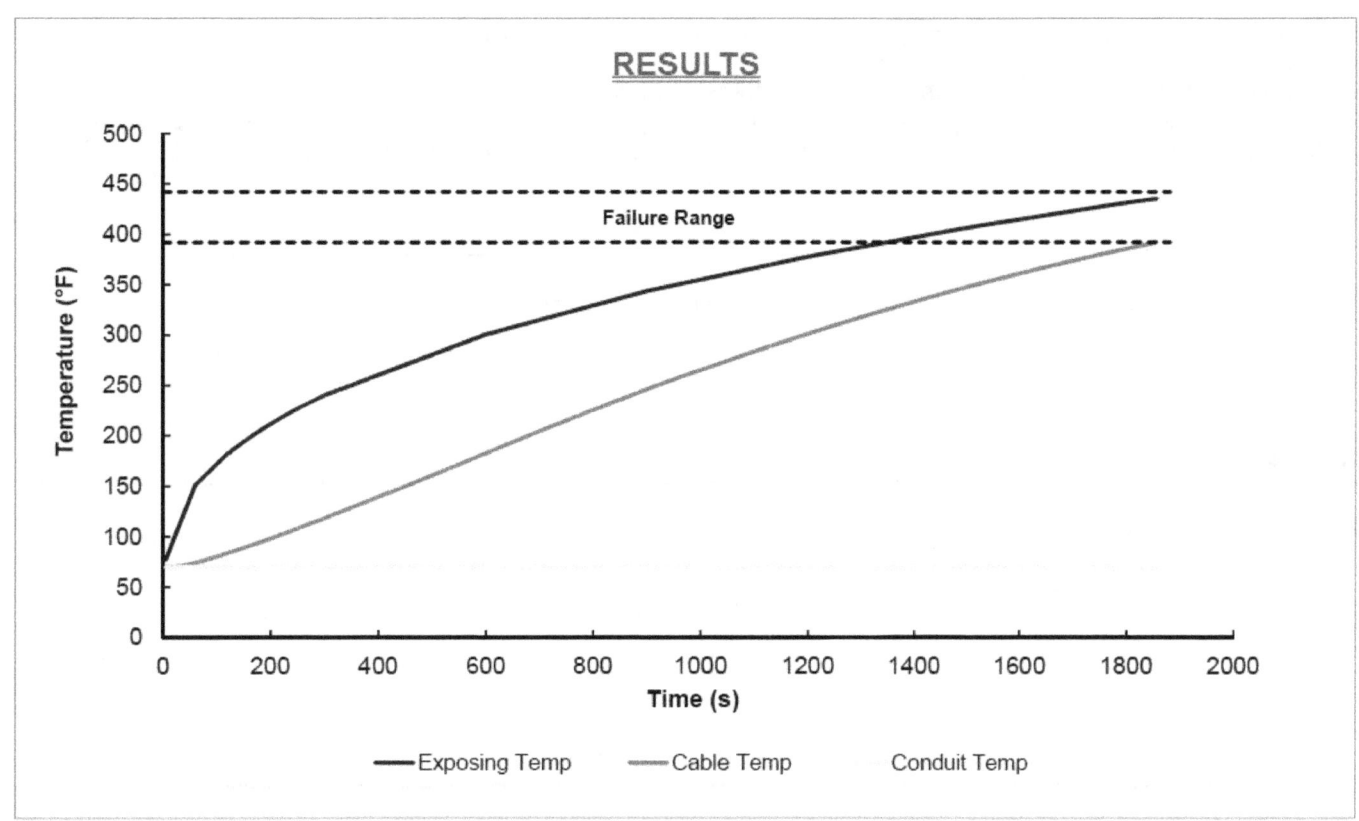

EXPOSURE GAS TEMPERATURE PROFILE		
Time (s)	Gas Temperature (°F)	Gas Temperature (°K)
0	70.00	294.26
60	150.98	339.25
120	182.16	356.57
180	205.22	369.38
240	224.11	379.88
300	240.35	388.90
600	300.66	422.41
900	343.58	446.25
1200	377.70	465.21
1500	406.33	481.11
1800	431.14	494.89
2100	453.10	507.09
2400	472.84	518.06
2700	490.81	528.04
3000	507.30	537.21
3300	522.57	545.69
3600	536.78	553.58

Answer: Cable fails at 30.9 minutes

COMPARTMENT WITH THERMALLY THICK/THIN BOUNDARIES

The following calculations estimate the hot gas layer temperature and smoke layer height in enclosure fire.
Parameters in YELLOW CELLS are Entered by the User.
Parameters in GREEN CELLS are Automatically Selected from the DROP DOWN MENU for the Material Selected.
All subsequent output values are calculated by the spreadsheet and based on values specified in the input parameters. This spreadsheet is protected and secure to avoid errors due to a wrong entry in a cell(s). The chapter in the NUREG should be read before an analysis is made.

Project / Inspection Title:	NUREG-1805 Supplement 1 Example 19-11.4c

INPUT PARAMETERS

COMPARTMENT INFORMATION

Compartment Width (w_c)	16.40	ft
Compartment Length (l_c)	16.40	ft
Compartment Height (h_c)	11.48	ft
Interior Lining Thickness (δ)	12.00	in

AMBIENT CONDITIONS

Ambient Air Temperature (T_a)	70.00	°F
Specific Heat of Air (c_p)	1.00	kJ/kg-K
Ambient Air Density (ρ_a)	1.20	kg/m^3

Note: Ambient Air Density (ρ_a) will automatically correct with Ambient Air Temperature (T_a) Input

THERMAL PROPERTIES OF COMPARTMENT ENCLOSING SURFACES FOR

Interior Lining Thermal Inertia ($k\rho c$)	2.9	(kW/m^2-K)^2-sec
Interior Lining Thermal Conductivity (k)	0.0016	kW/m-K
Interior Lining Specific Heat (c)	0.75	kJ/kg-K
Interior Lining Density (ρ)	2400	kg/m^3

THERMAL PROPERTIES FOR COMMON INTERIOR LINING MATERIALS

Material	kpc $(kW/m^2-K)^2$-sec	k (kW/m-K)	c (kJ/kg-K)	ρ (kg/m^3)	Select Material
					Concrete ▾
Aluminum (pure)	500	0.206	0.895	2710	**Scroll to desired material**
Steel (0.5% Carbon)	197	0.054	0.465	7850	**Click on selection**
Concrete	2.9	0.0016	0.75	2400	
Brick	1.7	0.0008	0.8	2600	
Glass, Plate	1.6	0.00076	0.8	2710	
Brick/Concrete Block	1.2	0.00073	0.84	1900	
Gypsum Board	0.18	0.00017	1.1	960	
Plywood	0.16	0.00012	2.5	540	
F ber Insulation Board	0.16	0.00053	1.25	240	
Chipboard	0.15	0.00015	1.25	800	
Aerated Concrete	0.12	0.00026	0.96	500	
Plasterboard	0.12	0.00016	0.84	950	
Calcium Silicate Board	0.098	0.00013	1.12	700	
Alumina Silicate Block	0.036	0.00014	1	260	
Glass Fiber Insulation	0.0018	0.000037	0.8	60	
Expanded Polystyrene	0.001	0.000034	1.5	20	
User Specified Value	Enter Value	Enter Value	Enter Value	Enter Value	

Reference: Klote, J., J. Milke, Principles of Smoke Management, 2002 Page 270.

COMPARTMENT MASS VENTILATION FLOW RATE

Forced Ventilation Flow Rate (m)   1000.00 cfm

FIRE SPECIFICATIONS

Fire Heat Release Rate (Q)   500.00 kW

**Click Here to Calculate this Sheet
and Return to THIEF**

Compartment Hot Gas Layer Temperature With Forced Ventilation

$$\Delta T_g / T_a = 0.63 (Q/mc_p T_a)^{0.72} (h_k A_T / mc_p)^{-0.36}$$

$$\Delta T_g = T_g - T_a$$

$$T_g = \Delta T_g + T_a$$

Results	Time After Ignition (t)		h_k (kW/m²-K)	$\Delta T_g/T_a$	ΔT_g (°K)	T_g (°K)	T_g (°C)	T_g (°F)
	(min)	(sec)						
	0	0	-	-	-	294.11	21.11	70.00
	1	60	0.22	0.35	102.58	396.69	123.69	254.64
	2	120	0.16	0.40	116.21	410.32	137.32	279.18
	3	180	0.13	0.43	125.01	419.12	146.12	295.01
	4	240	0.11	0.45	131.65	425.76	152.76	306.97
	5	300	0.10	0.47	137.05	431.16	158.16	316.69
	10	600	0.07	0.53	155.26	449.37	176.37	349.47
	15	900	0.06	0.57	167.01	461.12	188.12	370.62
	20	1200	0.05	0.60	175.89	470.00	197.00	386.60
	25	1500	0.04	0.62	183.10	477.21	204.21	399.58
	30	1800	0.04	0.64	189.21	483.32	210.32	410.57
	35	2100	0.04	0.66	194.53	488.64	215.64	420.16
	40	2400	0.03	0.68	199.26	493.37	220.37	428.67
	45	2700	0.03	0.69	203.53	497.64	224.64	436.36
	50	3000	0.03	0.71	207.43	501.54	228.54	443.37
	55	3300	0.03	0.72	211.02	505.13	232.13	449.83
	60	3600	0.03	0.73	214.35	508.46	235.46	455.83

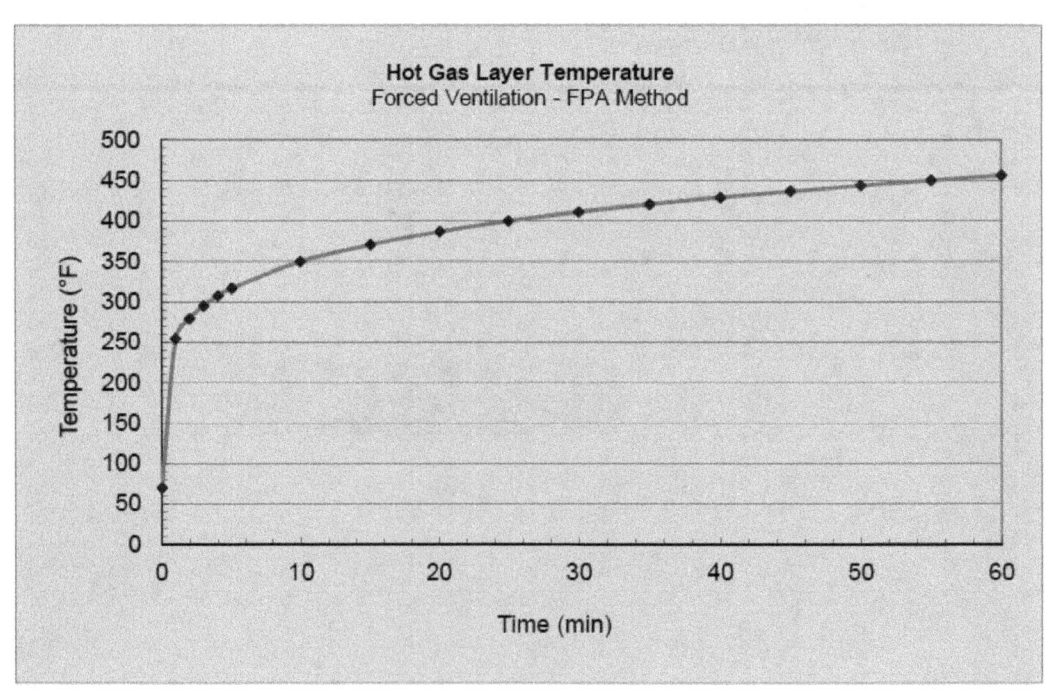

Hot Gas Layer Temperature
Forced Ventilation - FPA Method

Results	Time After Ignition (t)		h_k	ΔT_g	T_g	T_g	T_g
	(min)	(sec)	(kW/m²-K)	(°K)	(°K)	(°C)	(°F)
	0	0	-	-	294.11	21.11	70.00
	1	60	0.09	44.99	339.10	66.10	150.98
	2	120	0.06	62.31	356.42	83.42	182.16
	3	180	0.05	75.12	369.23	96.23	205.22
	4	240	0.04	85.61	379.73	106.73	224.11
	5	300	0.04	94.64	388.75	115.75	240.35
	10	600	0.03	128.15	422.26	149.26	300.66
	15	900	0.02	151.99	446.10	173.10	343.58
	20	1200	0.02	170.95	465.06	192.06	377.70
	25	1500	0.02	186.85	480.96	207.96	406.33
	30	1800	0.02	200.63	494.74	221.74	431.14
	35	2100	0.01	212.83	506.94	233.94	453.10
	40	2400	0.01	223.80	517.91	244.91	472.84
	45	2700	0.01	233.78	527.89	254.89	490.81
	50	3000	0.01	242.95	537.06	264.06	507.30
	55	3300	0.01	251.43	545.54	272.54	522.57
	60	3600	0.01	259.32	553.43	280.43	536.78

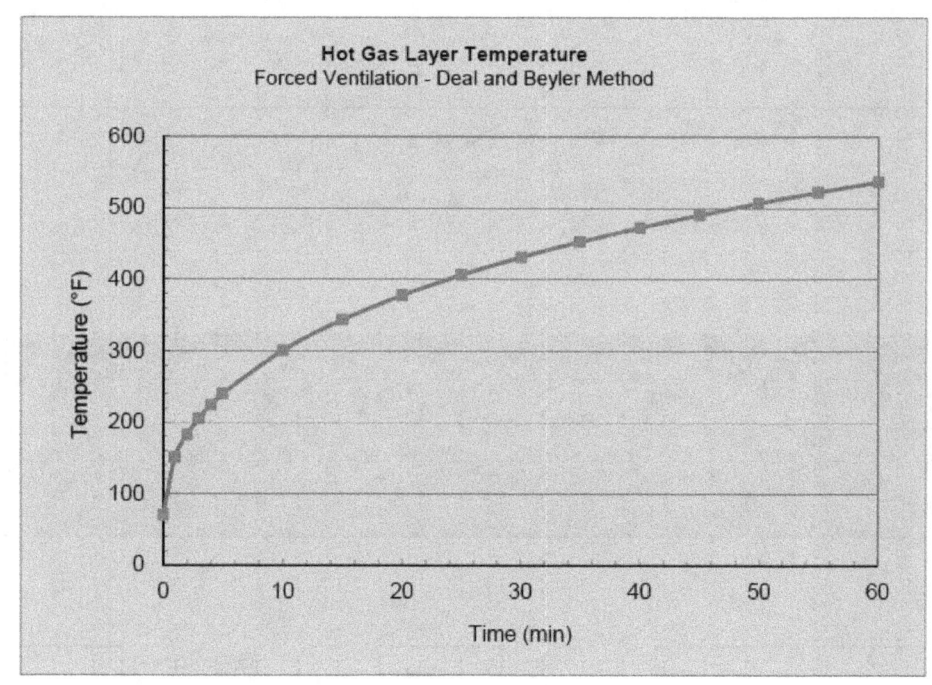

Hot Gas Layer Temperature
Forced Ventilation - Deal and Beyler Method

Summary of Results

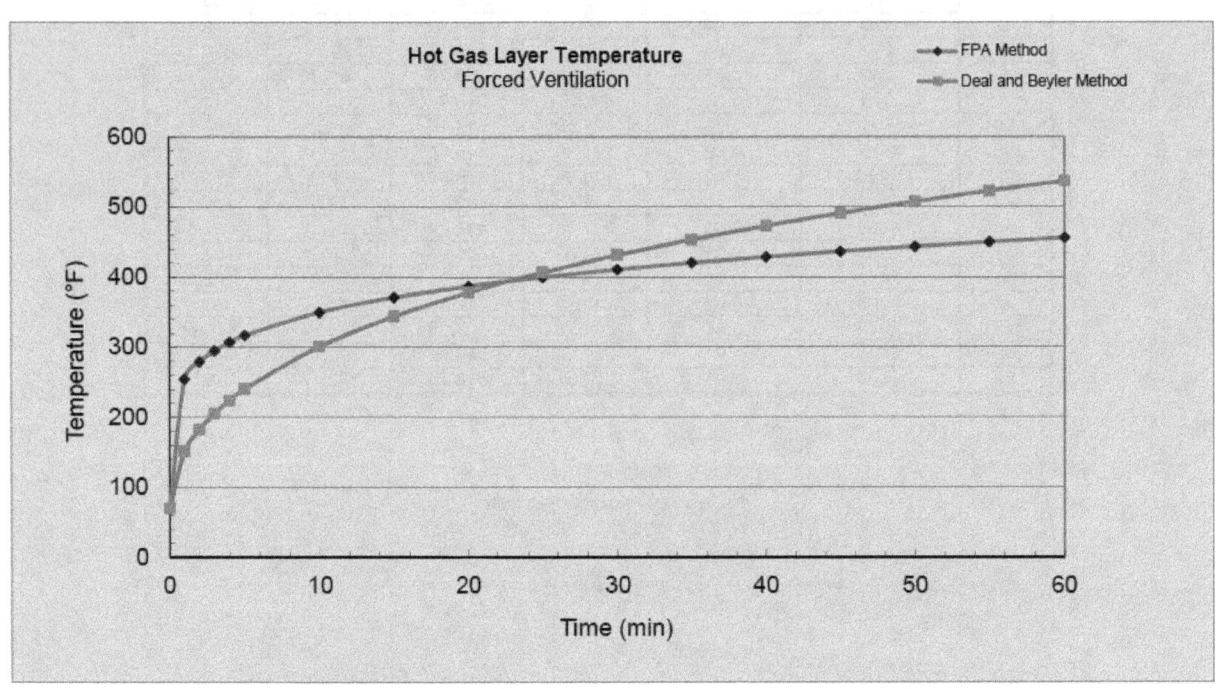

NOTE:
The above calculations are based on principles developed in the SFPE Handbook of Fire Protection Engineering, 2nd Edition, 1995. Calculations are based on certain assumptions and have inherent limitations. The results of such calculations may or may not have reasonable predictive capabilities for a given situation, and should only be interpreted by an informed user. Although each calculation in the spreadsheet has been verified with the results of hand calculation, there is no absolute guarantee of the accuracy of these calculations. Any questions, comments, concerns, and suggestions, or to report an error(s) in the spreadsheets, please send an email to David.Stroup@nrc.gov and Naeem.Iqbal@nrc.gov or MarkHenry.Salley@nrc.gov.

Prepared by: Date: Organization:

Checked by: Date: Organization:

Additional Information:

The following calculations estimate the time to failure of cables exposed to a specified hot gas layer.

Parameters in YELLOW CELLS are Entered by the User.

Parameters in GREEN CELLS are Automatically Selected from the DROP DOWN MENU or SELECT CABLE BUTTON.

All subsequent output values are calculated by the spreadsheet and based on values specified in the input parameters. This spreadsheet is protected and secure to avoid errors due to a wrong entry in a cell(s). The chapter in the NUREG should be read before an analysis is made.

Project / Inspection Title:	NUREG-1805 Supplement 1 Example 19.11-4c

INPUT PARAMETERS

Cable Diameter	0.62	in
Cable Mass per Unit Length	0.26	b/ft
Cable Jacket Thickness	0.060	in
Ambient Air Temperature	70	°F
Failure Temperature	752	°F
Maximum Time	4000	s
Conduit Thickness	0.00	in
Conduit Outside Diameter	0.00	in
Cable Density	1986.49	kg/m³
Cable Insulation Type (Thermoplastic or Thermoset)	Thermoset	
Cable Function (Control, Instrumentation or Power)	Control	
Wire Gauge (AWG)	14	
Number of Conductors	9	
Cable Location	Cable Tray	

Do Not Enter Any Values in the Green Boxes!

They are entered automatically based on the cable selection.

Select Cable

Click to Select Source for Exposure Gas Temperature Profile:

Natural Ventilation - Method of McCaffrey, Quintiere, Harkleroad (MQH)
Forced Ventilation - Method of Foote, Pagni, and Alvares (FPA)
Forced Ventilation - Method of Deal and Beyler
Room Fire with Closed Door
Within Fire Plume
User Defined

Selecting one of these items will automatically transfer you to the appropriate spreadsheet to calculate the exposure gas temperature profile

Warning: You MUST Click the Calculate Button Below when Finished Entering or Changing Data!

Calculate

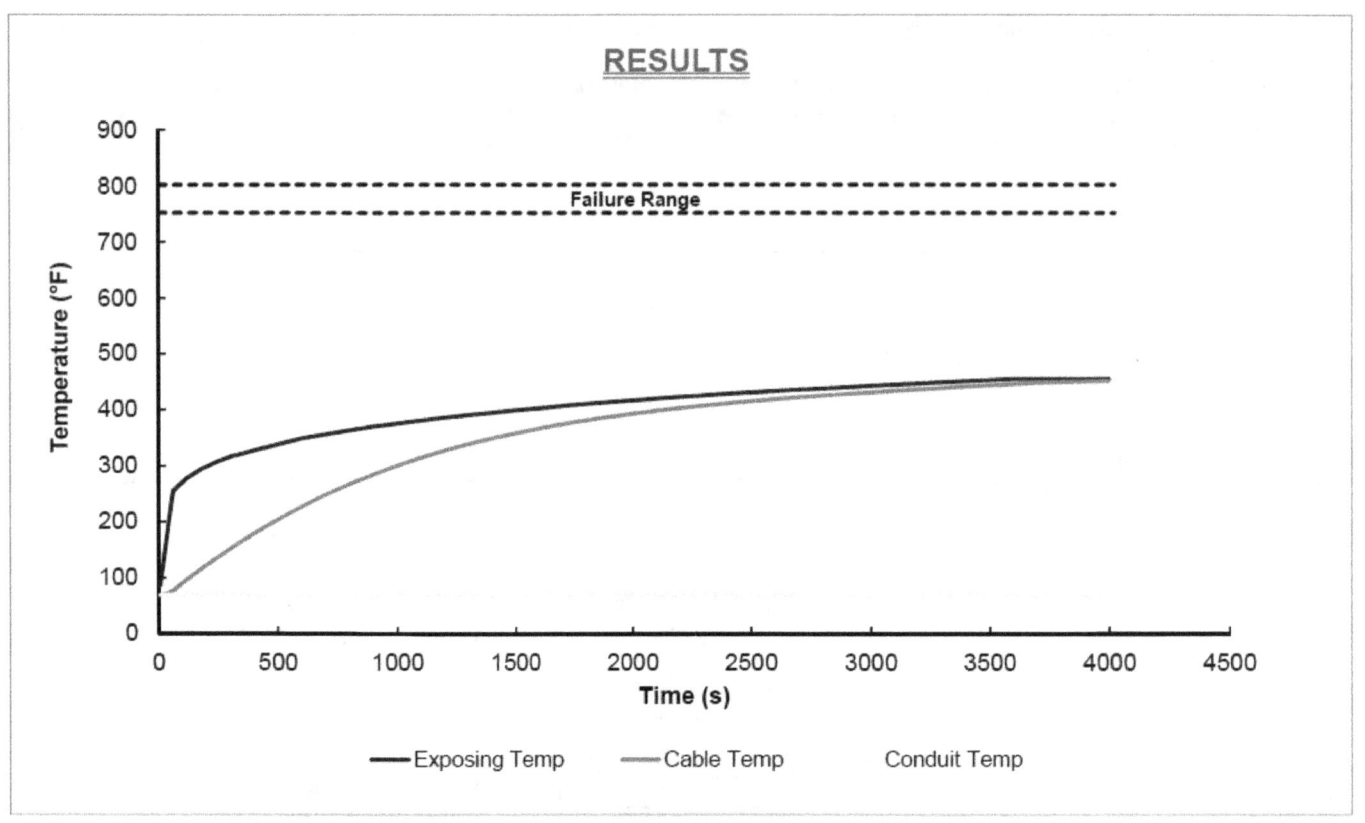

EXPOSURE GAS TEMPERATURE PROFILE		
Time (s)	Gas Temperature (°F)	Gas Temperature (°K)
0	70.00	294.26
60	254.64	396.84
120	279.18	410.47
180	295.01	419.27
240	306.97	425.91
300	316.69	431.31
600	349.47	449.52
900	370.62	461.27
1200	386.60	470.15
1500	399.58	477.36
1800	410.57	483.47
2100	420.16	488.79
2400	428.67	493.52
2700	436.36	497.79
3000	443.37	501.69
3300	449.83	505.28
3600	455.83	508.61

Answer: Cable does not reach failure temperature in 4000 seconds

COMPARTMENT WITH THERMALLY THICK/THIN BOUNDARIES

The following calculations estimate the hot gas layer temperature and smoke layer height in enclosure fire.

Parameters in YELLOW CELLS are Entered by the User.

Parameters in GREEN CELLS are Automatically Selected from the DROP DOWN MENU for the Material Selected.

All subsequent output values are calculated by the spreadsheet and based on values specified in the input parameters. This spreadsheet is protected and secure to avoid errors due to a wrong entry in a cell(s). The chapter in the NUREG should be read before an analysis is made.

Project / Inspection Title:	NUREG-1805 Supplement 1 Example 19-11.4d

INPUT PARAMETERS

COMPARTMENT INFORMATION

Compartment Width (w_c)	16.40	ft
Compartment Length (l_c)	16.40	ft
Compartment Height (h_c)	11.48	ft
Interior Lining Thickness (δ)	12.00	in

AMBIENT CONDITIONS

Ambient Air Temperature (T_a)	70.00	°F
Specific Heat of Air (c_p)	1.00	kJ/kg-K
Ambient Air Density (ρ_a)	1.20	kg/m³

Note: Ambient Air Density (ρ_a) will automatically correct with Ambient Air Temperature (T_a) Input

THERMAL PROPERTIES OF COMPARTMENT ENCLOSING SURFACES FOR

Interior Lining Thermal Inertia ($k\rho c$)	2.9	(kW/m²-K)²-sec
Interior Lining Thermal Conductivity (k)	0.0016	kW/m-K
Interior Lining Specific Heat (c)	0.75	kJ/kg-K
Interior Lining Density (ρ)	2400	kg/m³

THERMAL PROPERTIES FOR COMMON INTERIOR LINING MATERIALS

Material	kρc (kW/m²-K)²-sec	k (kW/m-K)	c (kJ/kg-K)	ρ (kg/m³)	Select Material
					Concrete ▼
Aluminum (pure)	500	0.206	0.895	2710	Scroll to desired material
Steel (0.5% Carbon)	197	0.054	0.465	7850	Click on selection
Concrete	2.9	0.0016	0.75	2400	
Brick	1.7	0.0008	0.8	2600	
Glass, Plate	1.6	0.00076	0.8	2710	
Brick/Concrete Block	1.2	0.00073	0.84	1900	
Gypsum Board	0.18	0.00017	1.1	960	
Plywood	0.16	0.00012	2.5	540	
F ber Insulation Board	0.16	0.00053	1.25	240	
Chipboard	0.15	0.00015	1.25	800	
Aerated Concrete	0.12	0.00026	0.96	500	
Plasterboard	0.12	0.00016	0.84	950	
Calcium Silicate Board	0.098	0.00013	1.12	700	
Alumina Silicate Block	0.036	0.00014	1	260	
Glass Fiber Insulation	0.0018	0.000037	0.8	60	
Expanded Polystyrene	0.001	0.000034	1.5	20	
User Specified Value	Enter Value	Enter Value	Enter Value	Enter Value	

Reference: Klote, J., J. Milke, Principles of Smoke Management, 2002 Page 270.

COMPARTMENT MASS VENTILATION FLOW RATE

Forced Ventilation Flow Rate (m)   1000.00 cfm

FIRE SPECIFICATIONS

Fire Heat Release Rate (Q)   500.00 kW

**Click Here to Calculate this Sheet
and Return to THIEF**

Compartment Hot Gas Layer Temperature With Forced Ventilation

$$\Delta T_g/T_a = 0.63(Q/mc_pT_a)^{0.72}(h_kA_T/mc_p)^{-0.36}$$

$$\Delta T_g = T_g - T_a$$

$$T_g = \Delta T_g + T_a$$

Results	Time After Ignition (t)		h_k	$\Delta T_g/T_a$	ΔT_g	T_g	T_g	T_g
	(min)	(sec)	(kW/m²-K)		(°K)	(°K)	(°C)	(°F)
	0	0	-	-	-	294.11	21.11	70.00
	1	60	0.22	0.35	102.58	396.69	123.69	254.64
	2	120	0.16	0.40	116.21	410.32	137.32	279.18
	3	180	0.13	0.43	125.01	419.12	146.12	295.01
	4	240	0.11	0.45	131.65	425.76	152.76	306.97
	5	300	0.10	0.47	137.05	431.16	158.16	316.69
	10	600	0.07	0.53	155.26	449.37	176.37	349.47
	15	900	0.06	0.57	167.01	461.12	188.12	370.62
	20	1200	0.05	0.60	175.89	470.00	197.00	386.60
	25	1500	0.04	0.62	183.10	477.21	204.21	399.58
	30	1800	0.04	0.64	189.21	483.32	210.32	410.57
	35	2100	0.04	0.66	194.53	488.64	215.64	420.16
	40	2400	0.03	0.68	199.26	493.37	220.37	428.67
	45	2700	0.03	0.69	203.53	497.64	224.64	436.36
	50	3000	0.03	0.71	207.43	501.54	228.54	443.37
	55	3300	0.03	0.72	211.02	505.13	232.13	449.83
	60	3600	0.03	0.73	214.35	508.46	235.46	455.83

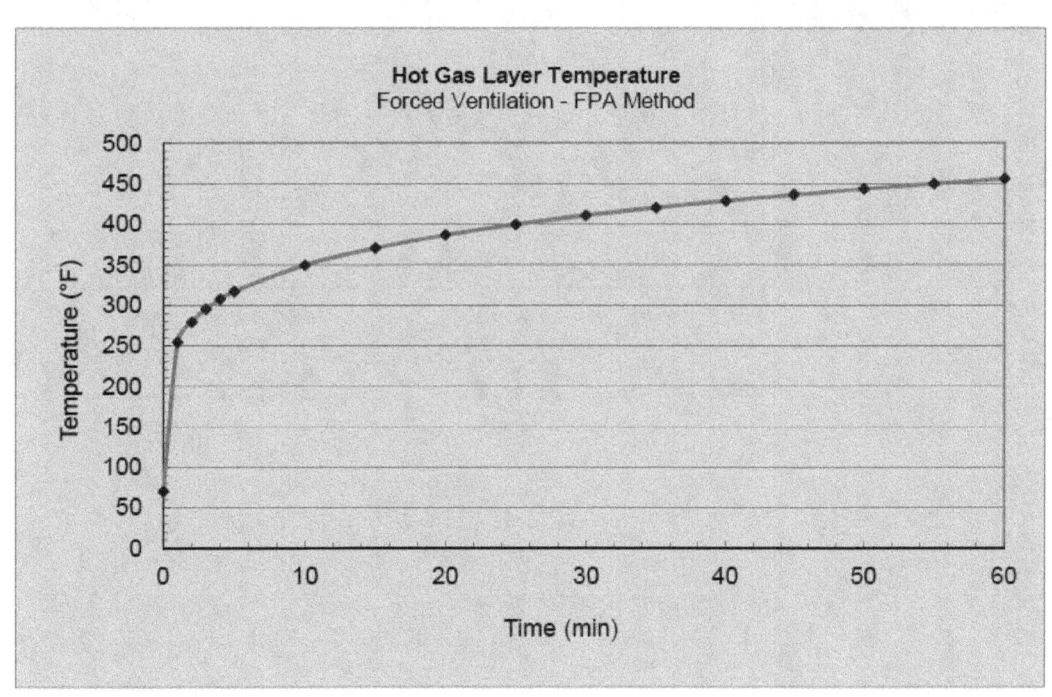

Hot Gas Layer Temperature
Forced Ventilation - FPA Method

Results	Time After Ignition (t)		h_k	ΔT_g	T_g	T_g	T_g
	(min)	(sec)	(kW/m²-K)	(°K)	(°K)	(°C)	(°F)
	0	0	-	-	294.11	21.11	70.00
	1	60	0.09	44.99	339.10	66.10	150.98
	2	120	0.06	62.31	356.42	83.42	182.16
	3	180	0.05	75.12	369.23	96.23	205.22
	4	240	0.04	85.61	379.73	106.73	224.11
	5	300	0.04	94.64	388.75	115.75	240.35
	10	600	0.03	128.15	422.26	149.26	300.66
	15	900	0.02	151.99	446.10	173.10	343.58
	20	1200	0.02	170.95	465.06	192.06	377.70
	25	1500	0.02	186.85	480.96	207.96	406.33
	30	1800	0.02	200.63	494.74	221.74	431.14
	35	2100	0.01	212.83	506.94	233.94	453.10
	40	2400	0.01	223.80	517.91	244.91	472.84
	45	2700	0.01	233.78	527.89	254.89	490.81
	50	3000	0.01	242.95	537.06	264.06	507.30
	55	3300	0.01	251.43	545.54	272.54	522.57
	60	3600	0.01	259.32	553.43	280.43	536.78

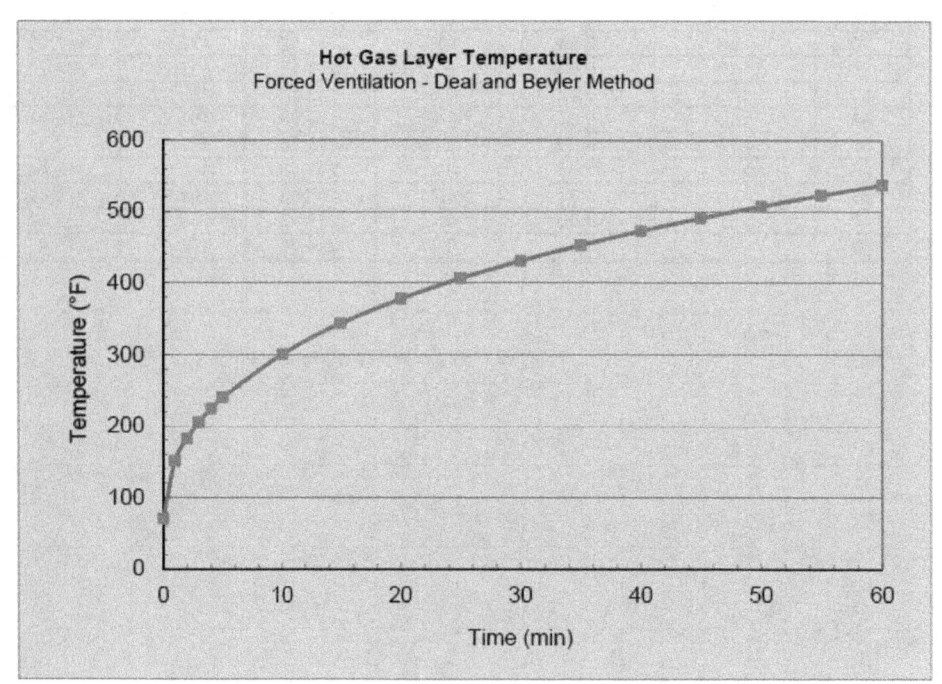

Hot Gas Layer Temperature
Forced Ventilation - Deal and Beyler Method

Summary of Results

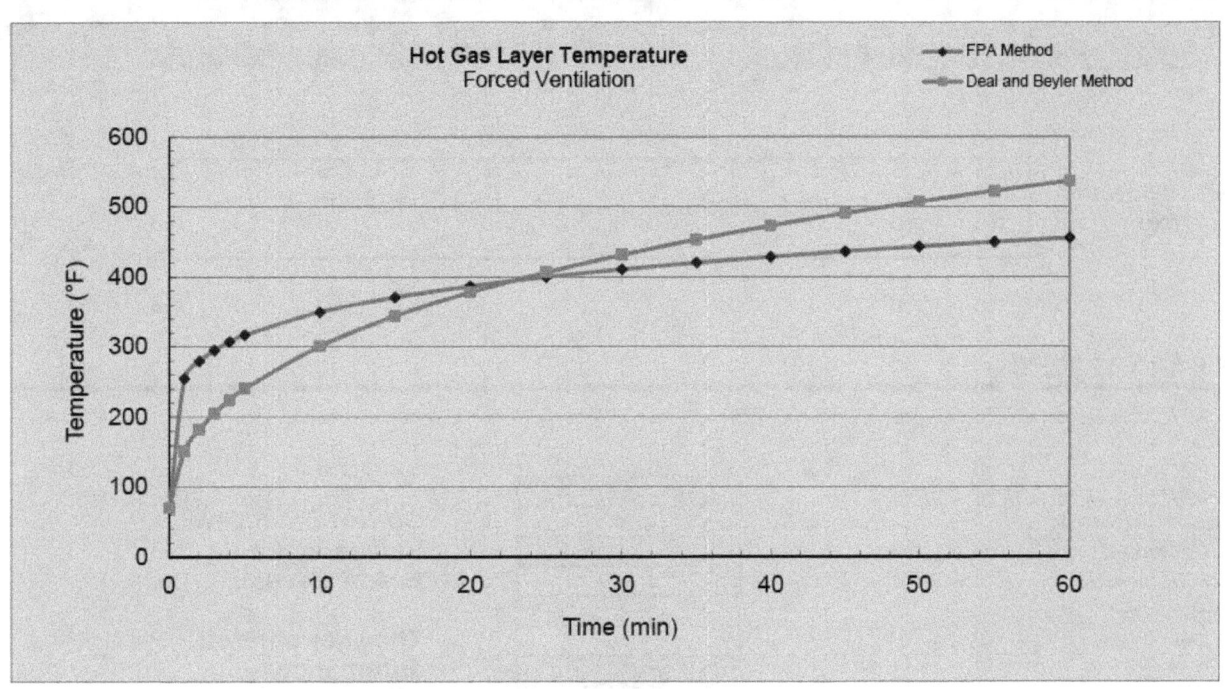

NOTE:
The above calculations are based on principles developed in the SFPE Handbook of Fire Protection Engineering, 2nd Edition, 1995. Calculations are based on certain assumptions and have inherent limitations. The results of such calculations may or may not have reasonable predictive capabilities for a given situation, and should only be interpreted by an informed user. Although each calculation in the spreadsheet has been verified with the results of hand calculation, there is no absolute guarantee of the accuracy of these calculations. Any questions, comments, concerns, and suggestions, or to report an error(s) in the spreadsheets, please send an email to David.Stroup@nrc.gov and Naeem.Iqbal@nrc.gov or MarkHenry.Salley@nrc.gov.

Prepared by: [   ] Date: [   ] Organization: [   ]

Checked by: [   ] Date: [   ] Organization: [   ]

Additional Information:

The following calculations estimate the time to failure of cables exposed to a specified hot gas layer.

Parameters in YELLOW CELLS are Entered by the User.

Parameters in GREEN CELLS are Automatically Selected from the DROP DOWN MENU or SELECT CABLE BUTTON.

All subsequent output values are calculated by the spreadsheet and based on values specified in the input parameters. This spreadsheet is protected and secure to avoid errors due to a wrong entry in a cell(s). The chapter in the NUREG should be read before an analysis is made.

Project / Inspection Title:	NUREG-1805 Supplement 1 Example 19.11-4d

INPUT PARAMETERS

Cable Diameter	0.62	in
Cable Mass per Unit Length	0.26	b/ft
Cable Jacket Thickness	0.060	in
Ambient Air Temperature	70	°F
Failure Temperature	752	°F
Maximum Time	4000	s
Conduit Thickness	0.00	in
Conduit Outside Diameter	0.00	in
Cable Density	1986.49	kg/m³
Cable Insulation Type (Thermoplastic or Thermoset)	Thermoset	
Cable Function (Control, Instrumentation or Power)	Control	
Wire Gauge (AWG)	14	
Number of Conductors	9	
Cable Location	Cable Tray	

Do Not Enter Any Values in the Green Boxes!

They are entered automatically based on the cable selection.

Select Cable

Click to Select Source for Exposure Gas Temperature Profile:

Natural Ventilation - Method of McCaffrey, Quintiere, Harkleroad (MQH)
Forced Ventilation - Method of Foote, Pagni, and Alvares (FPA)
Forced Ventilation - Method of Deal and Beyler
Room Fire with Closed Door
Within Fire Plume
User Defined

Selecting one of these items will automatically transfer you to the appropriate spreadsheet to calculate the exposure gas temperature profile

Warning: You MUST Click the Calculate Button Below when Finished Entering or Changing Data!

Calculate

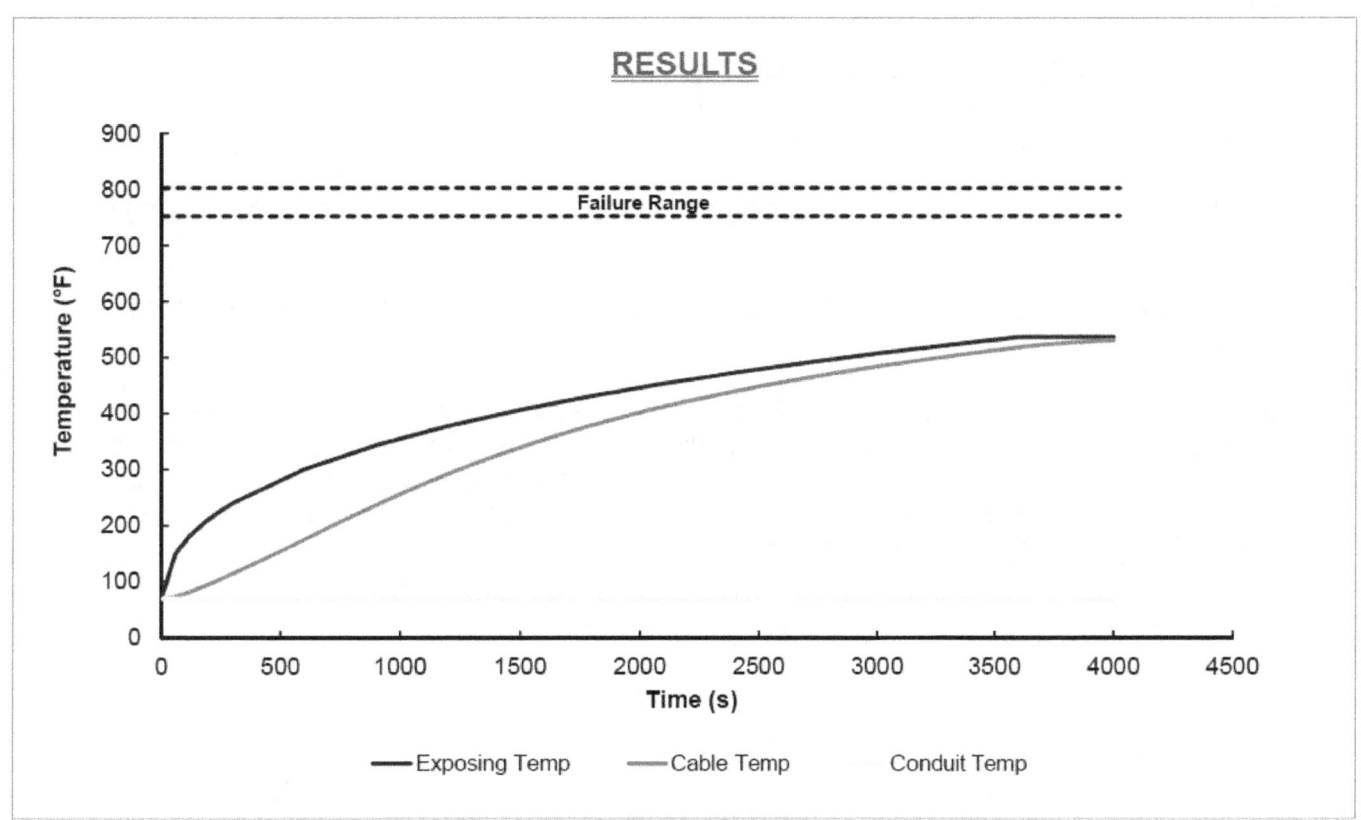

RESULTS

— Exposing Temp   — Cable Temp   Conduit Temp

EXPOSURE GAS TEMPERATURE PROFILE		
Time (s)	Gas Temperature (°F)	Gas Temperature (°K)
0	70.00	294.26
60	150.98	339.25
120	182.16	356.57
180	205.22	369.38
240	224.11	379.88
300	240.35	388.90
600	300.66	422.41
900	343.58	446.25
1200	377.70	465.21
1500	406.33	481.11
1800	431.14	494.89
2100	453.10	507.09
2400	472.84	518.06
2700	490.81	528.04
3000	507.30	537.21
3300	522.57	545.69
3600	536.78	553.58

Answer: Cable does not reach failure temperature in 4000 seconds

Example Problem 19.11-5 (SI Units)
(Cable Selection when Specific Cable is not on THIEF cable list)

Problem Statement
Consider the same concrete compartment used in the previous problem (Example Problem 19.11-4):

- 5.00 m (16.40 ft) wide x 5.00 m (16.40 ft) long x 3.50 m (11.48 ft) high (w_c x l_c x h_c)
- Force ventilation rate of 0.47 m³/s (1000 cfm)
- Fire is constant with a HRR of 500 kW.

The cable to be analyzed is located in a cable tray near the ceiling of the compartment and has the following identification:

- "Lui–Monninger Wire and Cable Company CONTROL CABLE 19/C 10 AWG PE/PVC 90°C Type TC"

This cable is not included in the THIEF list of cables; the cable manufacture has gone out of business with no records available online and the licensee doesn't have records available for this cable. Use the THIEF spreadsheet to calculate times to failure for "generic" cables with similar construction. It is also of interest to determine how much the time to failure can be increased by enclosing the cable in a conduit.

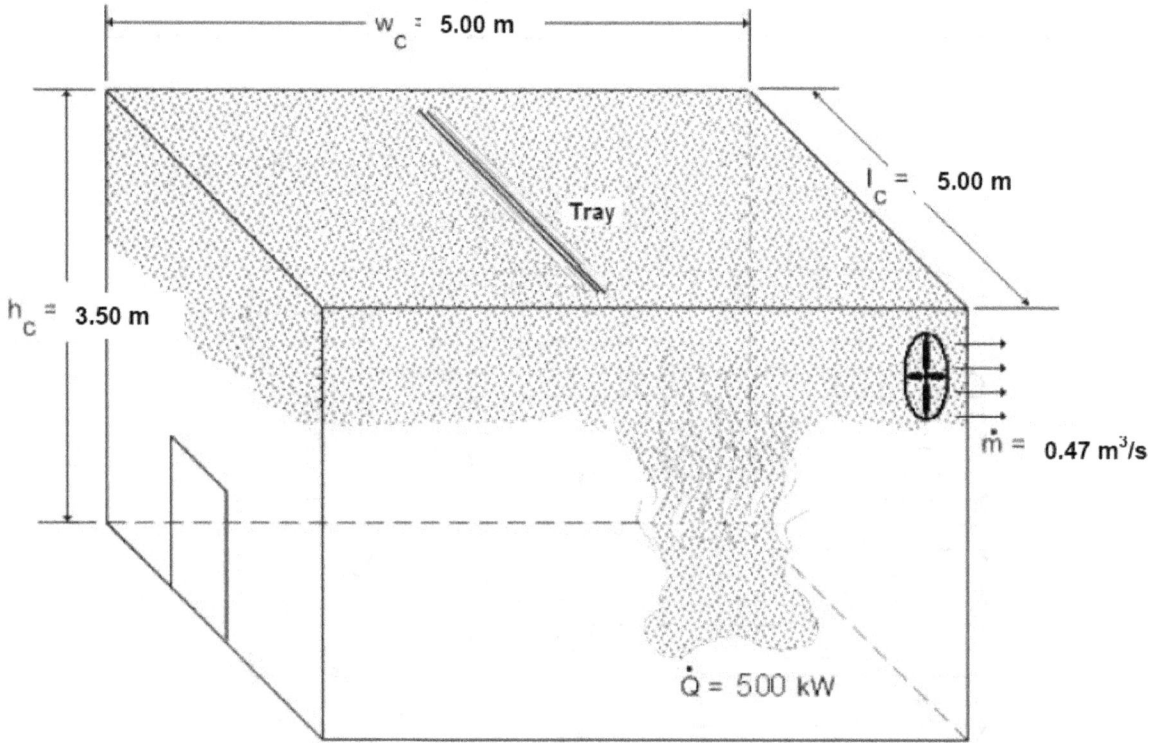

Example 19.11-5. Concrete Compartment with Forced Ventilation.

Solution

Purpose:

(1)   Determine the time to failure and compare the results of similar electrical cables located in a cable tray.

Assumptions:

(1)   The cable is surrounded by a uniform temperature hot gas layer.
(2)   The cable properties do not change with temperature increase.
(3)   Failure is indicated when the temperature inside the cable jacket exceeds 200 °C (392 °F) (thermoplastic cable).

Spreadsheet (FDT^s) Information:

Use the following FDT^s:

(a)   02.2_Temperature_FV_Sup1_SI.xls (click on *Temperature_FV*)
(b)   19_THIEF_Thermally_Induced_Electrical_Failure_of_Cables_Sup1_SI.xls (click on *THIEF*)

FDT^s Input Parameters:

-Gas temperature around the cable as a function time:

o   Use 02.2_Temperature_FV_Sup1_SI.xls and the parameters presented above to determine the gas temperature
o   On the THIEF spreadsheet, select: Forced Ventilation – Method of Deal and Beyler or Forced Ventilation – Method of Foote, Pagni, and Alvares (FPA)

-Press "Select Cable" Button
-Choose the "Generic" cable with TP insulation from the list for 10 AWG and 19 conductors
-Choose appropriate Cable Location ("Cable Tray") or 63 mm "Rigid"conduit
-Press "Calculate" Button

Results*

Cable Location	FPA Method (Minutes)	Deal and Beyler (Minutes)
Cable Tray	48.7	42.4
Conduit	No Failure	63

*see spreadsheets

As the results show, some variability exists between the different cable times to failure. A conservative approach would be to use the results of the cable that is most similar to the cable being analyzed (same insulation and jacket materials) and that has the shortest time to failure. In this example, cable 4 - General Cable – 20/10 – PE/PVC would be the best choice for representing the cable under analysis. The Deal and Beyler method results in a realistically conservative cable failure time of 42.4 minutes for the Lui-Monninger 10 AWG 19/C cable.

CHAPTER 2. PREDICTING HOT GAS LAYER TEMPERATURE
IN A ROOM FIRE
WITH FORCED VENTILATION

Version 1805.1
(SI Units)

COMPARTMENT WITH THERMALLY THICK/THIN BOUNDARIES

The following calculations estimate the hot gas layer temperature and smoke layer height in enclosure fire.
Parameters in YELLOW CELLS are Entered by the User.
Parameters in GREEN CELLS are Automatically Selected from the DROP DOWN MENU for the Material Selected.
All subsequent output values are calculated by the spreadsheet and based on values specified in the input parameters. This spreadsheet is protected and secure to avoid errors due to a wrong entry in a cell(s). The chapter in the NUREG should be read before an analysis is made.

Project / Inspection Title:	NUREG-1805 Supplement 1 Example 19.11-5a

INPUT PARAMETERS

COMPARTMENT INFORMATION

Compartment Width (w_c)	5.00	m
Compartment Length (l_c)	5.00	m
Compartment Height (h_c)	3.50	m
Interior Lining Thickness (δ)	30.00	cm

AMBIENT CONDITIONS

Ambient Air Temperature (T_a)	21.00	°C
Specific Heat of Air (c_p)	1.00	kJ/kg-K
Ambient Air Density (ρ_a)	1.20	kg/m^3

Note: Ambient Air Density (ρ_a) will automatically correct with Ambient Air Temperature (T_a) Input

THERMAL PROPERTIES OF COMPARTMENT ENCLOSING SURFACES FOR

Interior Lining Thermal Inertia ($k\rho c$)	2.9	(kW/m^2-K)^2-sec
Interior Lining Thermal Conductivity (k)	0.0016	kW/m-K
Interior Lining Specific Heat (c)	0.75	kJ/kg-K
Interior Lining Density (ρ)	2400	kg/m^3

THERMAL PROPERTIES FOR COMMON INTERIOR LINING MATERIALS

Material	kρc $(kW/m^2-K)^2$-sec	k (kW/m-K)	c (kJ/kg-K)	ρ (kg/m^3)	Select Material
					Concrete ▾
Aluminum (pure)	500	0.206	0.895	2710	**Scroll to desired material**
Steel (0.5% Carbon)	197	0.054	0.465	7850	**Click on selection**
Concrete	2.9	0.0016	0.75	2400	
Brick	1.7	0.0008	0.8	2600	
Glass, Plate	1.6	0.00076	0.8	2710	
Brick/Concrete Block	1.2	0.00073	0.84	1900	
Gypsum Board	0.18	0.00017	1.1	960	
Plywood	0.16	0.00012	2.5	540	
F ber Insulation Board	0.16	0.00053	1.25	240	
Chipboard	0.15	0.00015	1.25	800	
Aerated Concrete	0.12	0.00026	0.96	500	
Plasterboard	0.12	0.00016	0.84	950	
Calcium Silicate Board	0.098	0.00013	1.12	700	
Alumina Silicate Block	0.036	0.00014	1	260	
Glass Fiber Insulation	0.0018	0.000037	0.8	60	
Expanded Polystyrene	0.001	0.000034	1.5	20	
User Specified Value	Enter Value	Enter Value	Enter Value	Enter Value	

Reference: Klote, J., J. Milke, Principles of Smoke Management, 2002 Page 270.

COMPARTMENT MASS VENTILATION FLOW RATE

Forced Ventilation Flow Rate (m) | 0.47 | m^3/s

FIRE SPECIFICATIONS

Fire Heat Release Rate (Q) | 500.00 | kW

**Click Here to Calculate this
Sheet and Return to THIEF**

Compartment Hot Gas Layer Temperature With Forced Ventilation

$$\Delta T_g/T_a = 0.63(Q/mc_pT_a)^{0.72}(h_kA_T/mc_p)^{-0.36}$$

$$\Delta T_g = T_g - T_a$$

$$T_g = \Delta T_g + T_a$$

Results	Time After Ignition (t)		h_k	$\Delta T_g/T_a$	ΔT_g	T_g	T_g	T_g
	(min)	(sec)	(kW/m^2-K)		(°K)	(°K)	(°C)	(°F)
	0	0	-	-	-	294.00	21.00	69.80
	1	60	0.22	0.35	102.69	396.69	123.69	254.64
	2	120	0.16	0.40	116.33	410.33	137.33	279.20
	3	180	0.13	0.43	125.14	419.14	146.14	295.05
	4	240	0.11	0.45	131.79	425.79	152.79	307.03
	5	300	0.10	0.47	137.19	431.19	158.19	316.75
	10	600	0.07	0.53	155.42	449.42	176.42	349.56
	15	900	0.06	0.57	167.19	461.19	188.19	370.74
	20	1200	0.05	0.60	176.08	470.08	197.08	386.74
	25	1500	0.04	0.62	183.29	477.29	204.29	399.73
	30	1800	0.04	0.64	189.41	483.41	210.41	410.74
	35	2100	0.04	0.66	194.74	488.74	215.74	420.33
	40	2400	0.03	0.68	199.47	493.47	220.47	428.85
	45	2700	0.03	0.69	203.75	497.75	224.75	436.55
	50	3000	0.03	0.71	207.65	501.65	228.65	443.57
	55	3300	0.03	0.72	211.24	505.24	232.24	450.04
	60	3600	0.03	0.73	214.58	508.58	235.58	456.04

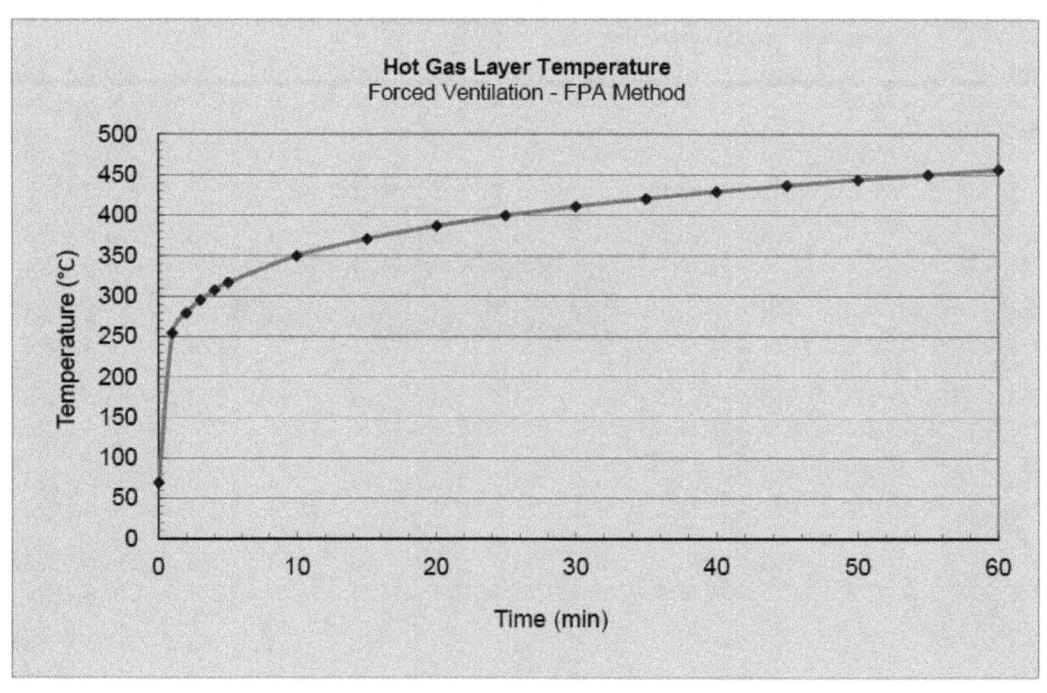

Hot Gas Layer Temperature
Forced Ventilation - FPA Method

Results	Time After Ignition (t)		h_k	ΔT_g	T_g	T_g	T_g
	(min)	(sec)	(kW/m²-K)	(°K)	(°K)	(°C)	(°F)
	0	0	-	-	294.00	21.00	69.80
	1	60	0.09	44.98	338.98	65.98	150.76
	2	120	0.06	62.30	356.30	83.30	181.93
	3	180	0.05	75.11	369.11	96.11	205.00
	4	240	0.04	85.61	379.61	106.61	223.89
	5	300	0.04	94.63	388.63	115.63	240.14
	10	600	0.03	128.16	422.16	149.16	300.49
	15	900	0.02	152.02	446.02	173.02	343.44
	20	1200	0.02	171.00	465.00	192.00	377.60
	25	1500	0.02	186.93	480.93	207.93	406.27
	30	1800	0.02	200.72	494.72	221.72	431.10
	35	2100	0.01	212.94	506.94	233.94	453.10
	40	2400	0.01	223.93	517.93	244.93	472.87
	45	2700	0.01	233.93	527.93	254.93	490.87
	50	3000	0.01	243.11	537.11	264.11	507.39
	55	3300	0.01	251.60	545.60	272.60	522.69
	60	3600	0.01	259.52	553.52	280.52	536.93

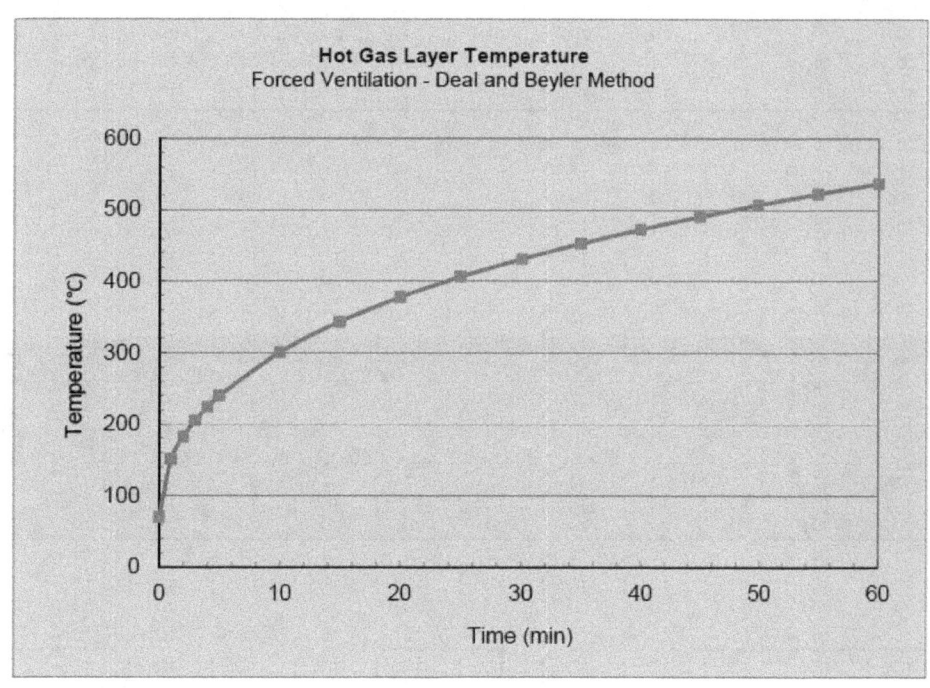

Hot Gas Layer Temperature
Forced Ventilation - Deal and Beyler Method

CHAPTER 2. PREDICTING HOT GAS LAYER TEMPERATURE IN A ROOM FIRE WITH FORCED VENTILATION

Version 1805.1
(SI Units)

Summary of Results

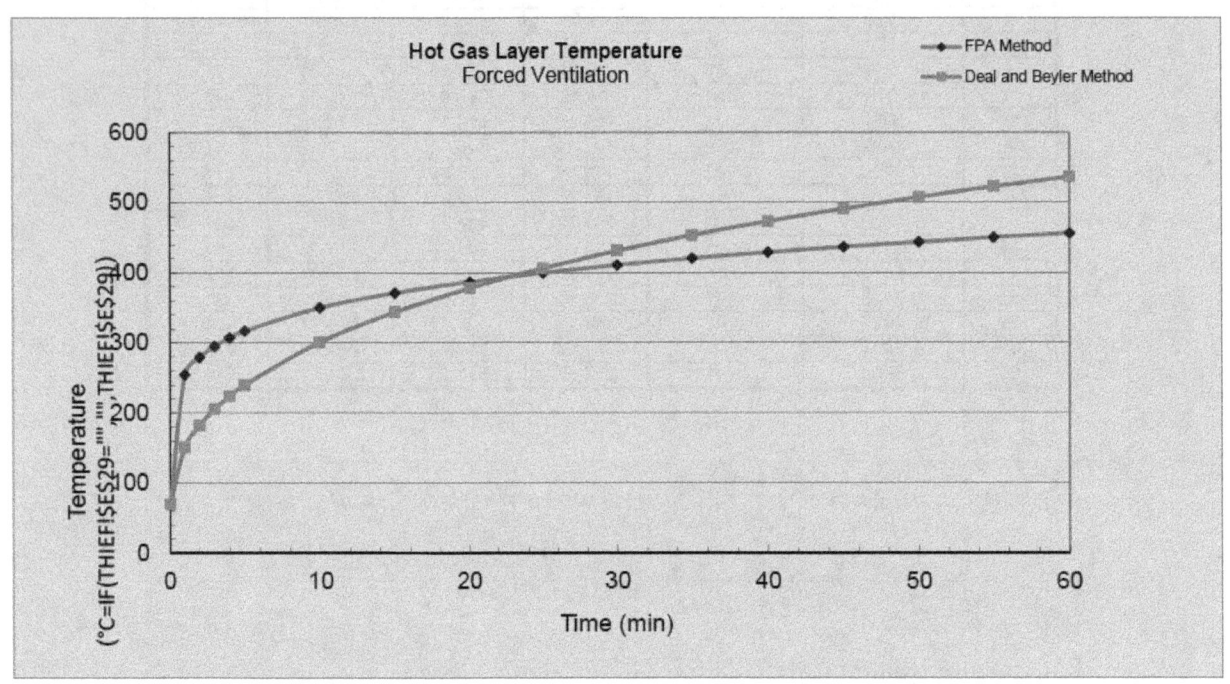

NOTE:

The above calculations are based on principles developed in the SFPE Handbook of Fire Protection Engineering, 2nd Edition, 1995. Calculations are based on certain assumptions and have inherent limitations. The results of such calculations may or may not have reasonable predictive capabilities for a given situation, and should only be interpreted by an informed user. Although each calculation in the spreadsheet has been verified with the results of hand calculation, there is no absolute guarantee of the accuracy of these calculations. Any questions, comments, concerns, and suggestions, or to report an error(s) in the spreadsheets, please send an email to David.Stroup@nrc.gov and Naeem.Iqbal@nrc.gov or MarkHenry.Salley@nrc.gov.

Prepared by: [ ]   Date: [ ]   Organization: [ ]

Checked by: [ ]   Date: [ ]   Organization: [ ]

Additional Information:

The following calculations estimate the time to failure of cables exposed to a specified hot gas layer.

Parameters in YELLOW CELLS are Entered by the User.

Parameters in GREEN CELLS are Automatically Selected from the DROP DOWN MENU or SELECT CABLE BUTTON.

All subsequent output values are calculated by the spreadsheet and based on values specified in the input parameters. This spreadsheet is protected and secure to avoid errors due to a wrong entry in a cell(s). The chapter in the NUREG should be read before an analysis is made.

Project / Inspection Title:	NUREG-1805 Supplement 1 Example 19.11-5a

INPUT PARAMETERS

Cable Diameter	26.67	mm
Cable Mass per Unit Length	1.38	kg/m
Cable Jacket Thickness	1.524	mm
Ambient Air Temperature	21	°C
Failure Temperature	200	°C
Maximum Time	4000	s
Conduit Thickness	0.00	mm
Conduit Outside Diameter	0.00	mm
Cable Density	2477.41	kg/m³
Cable Insulation Type (Thermoplastic or Thermoset)	Thermoplastic	
Cable Function (Control, Instrumentation or Power)	Control	
Wire Gauge (AWG)	10	
Number of Conductors	19	
Cable Location	Cable Tray	

Do Not Enter Any Values in the Green Boxes!

They are entered automatically based on the cable selection.

Select Cable

Click to Select Source for Exposure Gas Temperature Profile:

Natural Ventilation - Method of McCaffrey, Quintiere, Harkleroad (MQH)
Forced Ventilation - Method of Foote, Pagni, and Alvares (FPA)
Forced Ventilation - Method of Deal and Beyler
Room Fire with Closed Door
Within Fire Plume
User Defined

Selecting one of these items will automatically transfer you to the appropriate spreadsheet to calculate the exposure gas temperature profile

Warning: You MUST Click the Calculate Button Below when Finished Entering or Changing Data!

Calculate

RESULTS

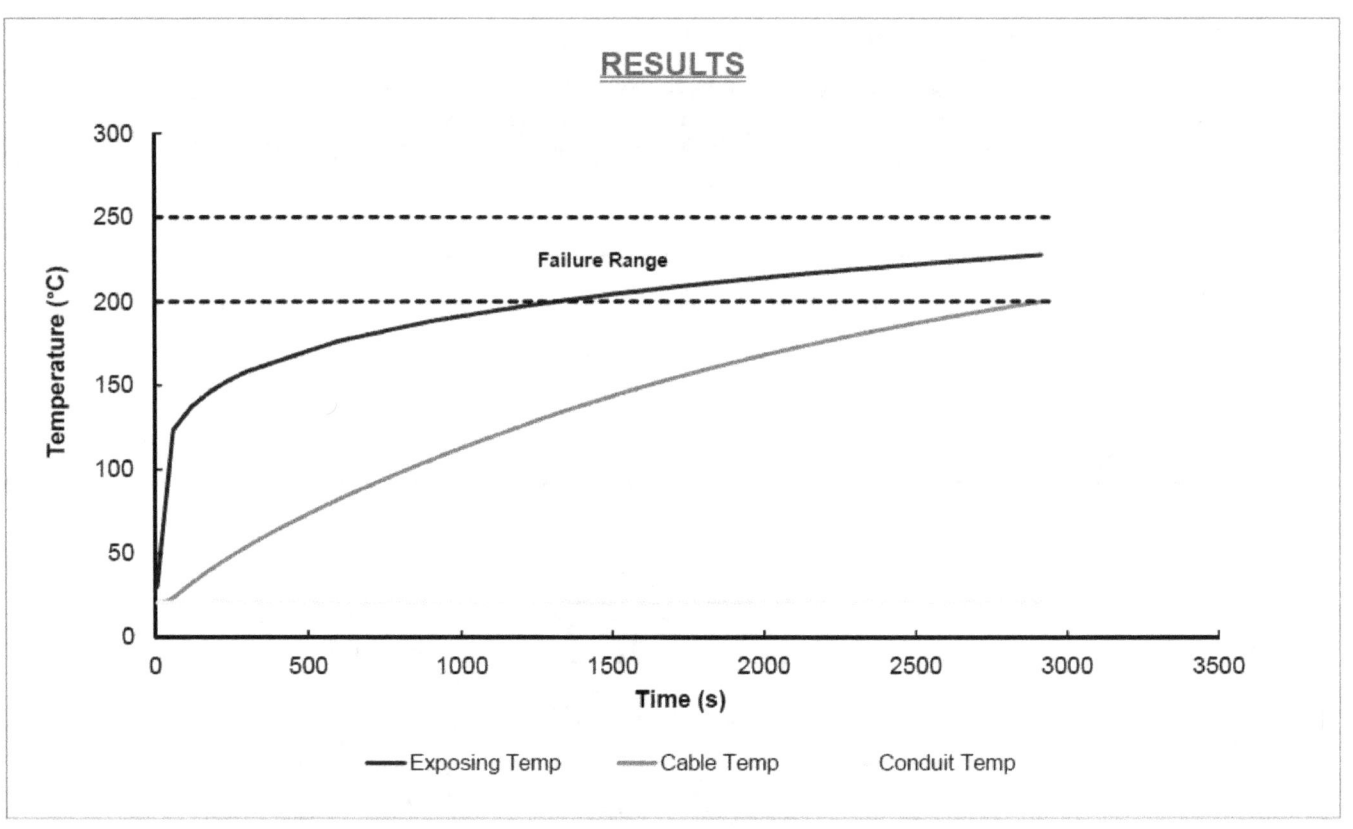

EXPOSURE GAS TEMPERATURE PROFILE		
Time (s)	Gas Temperature (°C)	Gas Temperature (°K)
0	21.00	294.15
60	123.69	396.84
120	137.33	410.48
180	146.14	419.29
240	152.79	425.94
300	158.19	431.34
600	176.42	449.57
900	188.19	461.34
1200	197.08	470.23
1500	204.29	477.44
1800	210.41	483.56
2100	215.74	488.89
2400	220.47	493.62
2700	224.75	497.90
3000	228.65	501.80
3300	232.24	505.39
3600	235.58	508.73

Answer: Cable fails at 48.7 minutes

COMPARTMENT WITH THERMALLY THICK/THIN BOUNDARIES

The following calculations estimate the hot gas layer temperature and smoke layer height in enclosure fire.

Parameters in YELLOW CELLS are Entered by the User.

Parameters in GREEN CELLS are Automatically Selected from the DROP DOWN MENU for the Material Selected.

All subsequent output values are calculated by the spreadsheet and based on values specified in the input parameters. This spreadsheet is protected and secure to avoid errors due to a wrong entry in a cell(s). The chapter in the NUREG should be read before an analysis is made.

Project / Inspection Title:	NUREG-1805 Supplement 1 Example 19.11-5b

INPUT PARAMETERS

COMPARTMENT INFORMATION

Compartment Width (w_c)	5.00	m
Compartment Length (l_c)	5.00	m
Compartment Height (h_c)	3.50	m
Interior Lining Thickness (δ)	30.00	cm

AMBIENT CONDITIONS

Ambient Air Temperature (T_a)	21.00	°C
Specific Heat of Air (c_p)	1.00	kJ/kg-K
Ambient Air Density (ρ_a)	1.20	kg/m³

Note: Ambient Air Density (ρ_a) will automatically correct with Ambient Air Temperature (T_a) Input

THERMAL PROPERTIES OF COMPARTMENT ENCLOSING SURFACES FOR

Interior Lining Thermal Inertia ($k\rho c$)	2.9	(kW/m²-K)²-sec
Interior Lining Thermal Conductivity (k)	0.0016	kW/m-K
Interior Lining Specific Heat (c)	0.75	kJ/kg-K
Interior Lining Density (ρ)	2400	kg/m³

THERMAL PROPERTIES FOR COMMON INTERIOR LINING MATERIALS

Material	kρc (kW/m²-K)²-sec	k (kW/m-K)	c (kJ/kg-K)	ρ (kg/m³)	Select Material
					Concrete ▼
Aluminum (pure)	500	0.206	0.895	2710	Scroll to desired material
Steel (0.5% Carbon)	197	0.054	0.465	7850	Click on selection
Concrete	2.9	0.0016	0.75	2400	
Brick	1.7	0.0008	0.8	2600	
Glass, Plate	1.6	0.00076	0.8	2710	
Brick/Concrete Block	1.2	0.00073	0.84	1900	
Gypsum Board	0.18	0.00017	1.1	960	
Plywood	0.16	0.00012	2.5	540	
F ber Insulation Board	0.16	0.00053	1.25	240	
Chipboard	0.15	0.00015	1.25	800	
Aerated Concrete	0.12	0.00026	0.96	500	
Plasterboard	0.12	0.00016	0.84	950	
Calcium Silicate Board	0.098	0.00013	1.12	700	
Alumina Silicate Block	0.036	0.00014	1	260	
Glass Fiber Insulation	0.0018	0.000037	0.8	60	
Expanded Polystyrene	0.001	0.000034	1.5	20	
User Specified Value	Enter Value	Enter Value	Enter Value	Enter Value	

Reference: Klote, J., J. Milke, Principles of Smoke Management, 2002 Page 270.

COMPARTMENT MASS VENTILATION FLOW RATE

Forced Ventilation Flow Rate (m)   0.47 m³/s

FIRE SPECIFICATIONS

Fire Heat Release Rate (Q)   500.00 kW

**Click Here to Calculate this
Sheet and Return to THIEF**

CHAPTER 2. PREDICTING HOT GAS LAYER TEMPERATURE
IN A ROOM FIRE
WITH FORCED VENTILATION

Version 1805.1
(SI Units)

Compartment Hot Gas Layer Temperature With Forced Ventilation

$$\Delta T_g / T_a = 0.63 (Q/mc_p T_a)^{0.72} (h_k A_T / mc_p)^{-0.36}$$

$$\Delta T_g = T_g - T_a$$

$$T_g = \Delta T_g + T_a$$

Results	Time After Ignition (t)		h_k	$\Delta T_g / T_a$	ΔT_g	T_g	T_g	T_g
	(min)	(sec)	(kW/m²-K)		(°K)	(°K)	(°C)	(°F)
	0	0	-	-	-	294.00	21.00	69.80
	1	60	0.22	0.35	102.69	396.69	123.69	254.64
	2	120	0.16	0.40	116.33	410.33	137.33	279.20
	3	180	0.13	0.43	125.14	419.14	146.14	295.05
	4	240	0.11	0.45	131.79	425.79	152.79	307.03
	5	300	0.10	0.47	137.19	431.19	158.19	316.75
	10	600	0.07	0.53	155.42	449.42	176.42	349.56
	15	900	0.06	0.57	167.19	461.19	188.19	370.74
	20	1200	0.05	0.60	176.08	470.08	197.08	386.74
	25	1500	0.04	0.62	183.29	477.29	204.29	399.73
	30	1800	0.04	0.64	189.41	483.41	210.41	410.74
	35	2100	0.04	0.66	194.74	488.74	215.74	420.33
	40	2400	0.03	0.68	199.47	493.47	220.47	428.85
	45	2700	0.03	0.69	203.75	497.75	224.75	436.55
	50	3000	0.03	0.71	207.65	501.65	228.65	443.57
	55	3300	0.03	0.72	211.24	505.24	232.24	450.04
	60	3600	0.03	0.73	214.58	508.58	235.58	456.04

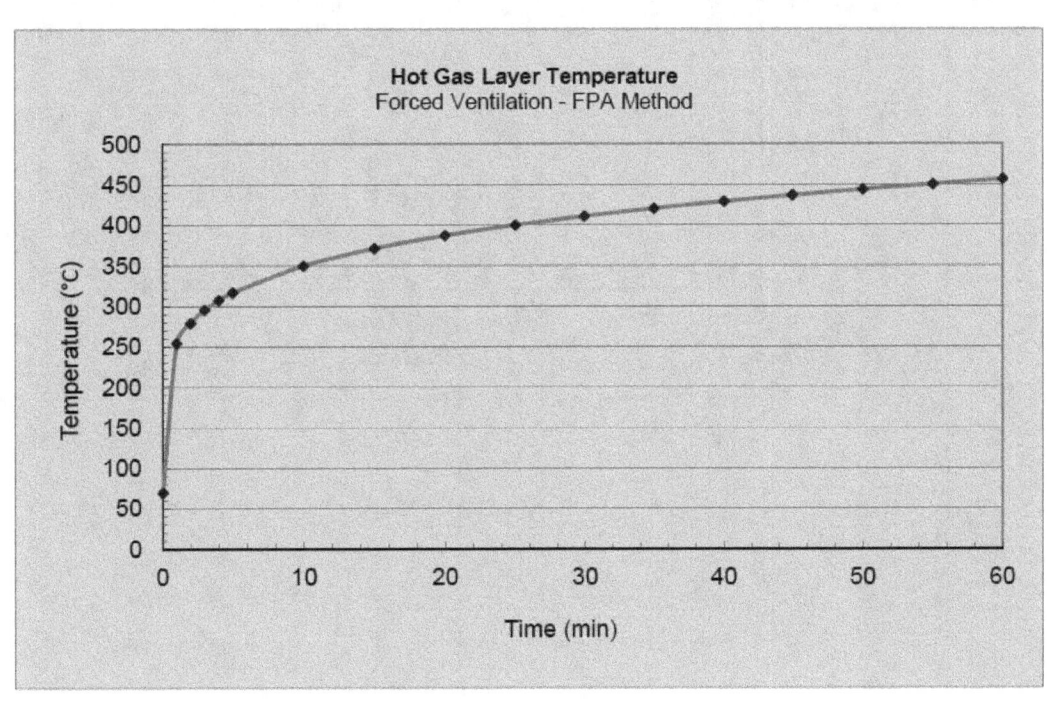

19-89

CHAPTER 2. PREDICTING HOT GAS LAYER TEMPERATURE
IN A ROOM FIRE
WITH FORCED VENTILATION

Version 1805.1
(SI Units)

Results	Time After Ignition (t)		h_k	ΔT_g	T_g	T_g	T_g
	(min)	(sec)	(kW/m²-K)	(°K)	(°K)	(°C)	(°F)
	0	0	-	-	294.00	21.00	69.80
	1	60	0.09	44.98	338.98	65.98	150.76
	2	120	0.06	62.30	356.30	83.30	181.93
	3	180	0.05	75.11	369.11	96.11	205.00
	4	240	0.04	85.61	379.61	106.61	223.89
	5	300	0.04	94.63	388.63	115.63	240.14
	10	600	0.03	128.16	422.16	149.16	300.49
	15	900	0.02	152.02	446.02	173.02	343.44
	20	1200	0.02	171.00	465.00	192.00	377.60
	25	1500	0.02	186.93	480.93	207.93	406.27
	30	1800	0.02	200.72	494.72	221.72	431.10
	35	2100	0.01	212.94	506.94	233.94	453.10
	40	2400	0.01	223.93	517.93	244.93	472.87
	45	2700	0.01	233.93	527.93	254.93	490.87
	50	3000	0.01	243.11	537.11	264.11	507.39
	55	3300	0.01	251.60	545.60	272.60	522.69
	60	3600	0.01	259.52	553.52	280.52	536.93

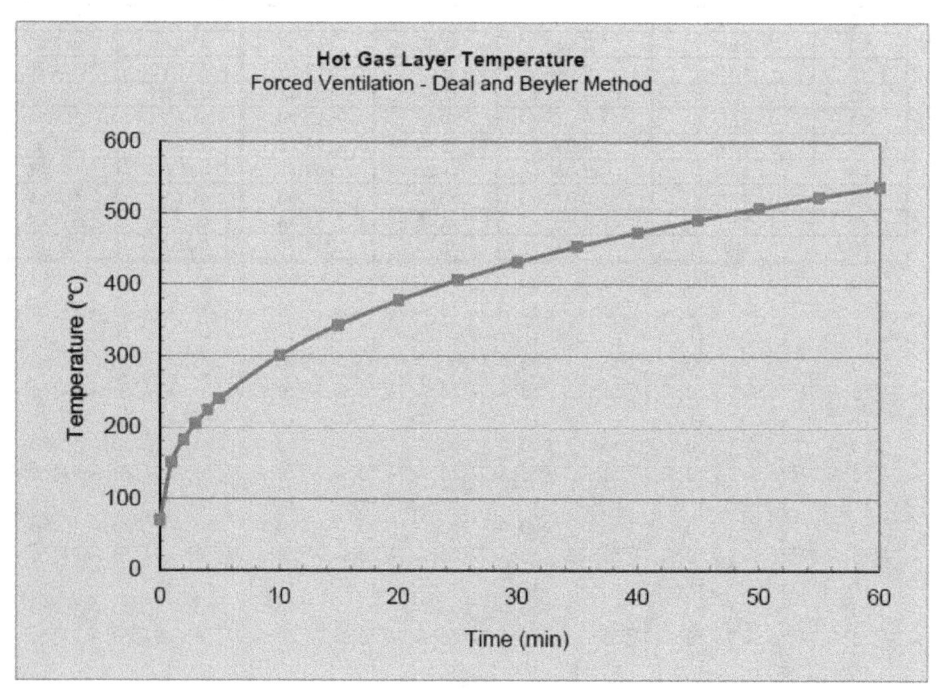

Hot Gas Layer Temperature
Forced Ventilation - Deal and Beyler Method

Summary of Results

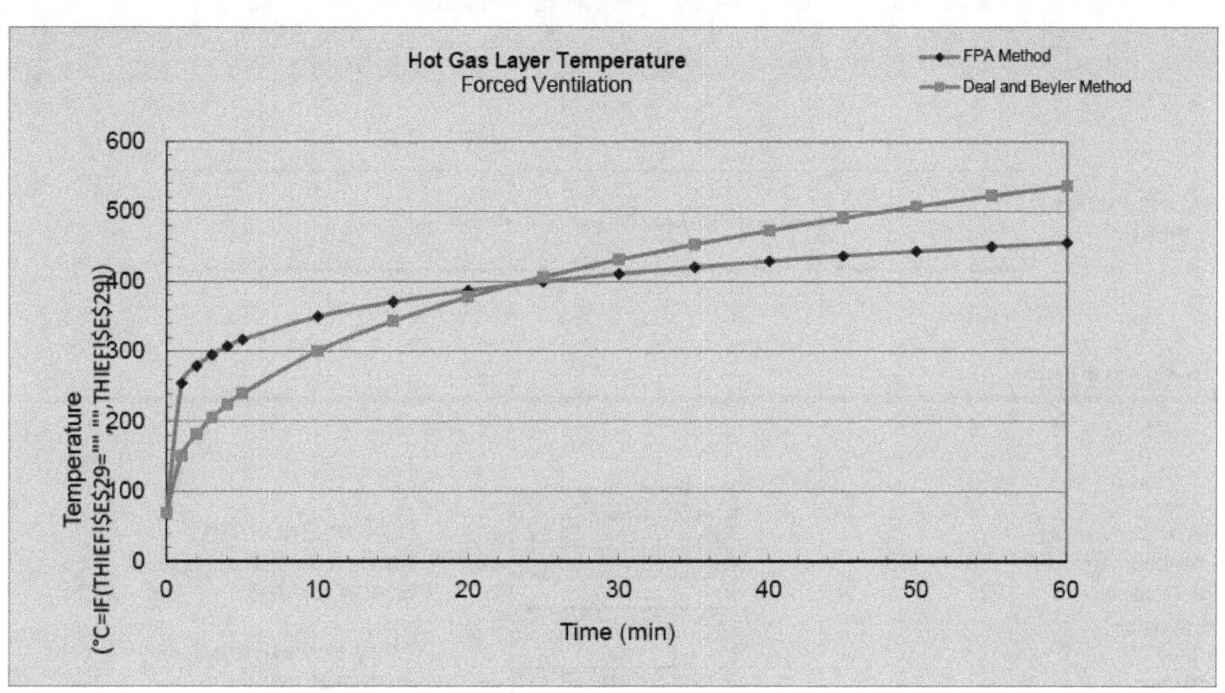

NOTE:

The above calculations are based on principles developed in the SFPE Handbook of Fire Protection Engineering, 2nd Edition, 1995. Calculations are based on certain assumptions and have inherent limitations. The results of such calculations may or may not have reasonable predictive capabilities for a given situation, and should only be interpreted by an informed user. Although each calculation in the spreadsheet has been verified with the results of hand calculation, there is no absolute guarantee of the accuracy of these calculations. Any questions, comments, concerns, and suggestions, or to report an error(s) in the spreadsheets, please send an email to David.Stroup@nrc.gov and Naeem.Iqbal@nrc.gov or MarkHenry.Salley@nrc.gov.

Prepared by: [ ] Date: [ ] Organization: [ ]

Checked by: [ ] Date: [ ] Organization: [ ]

Additional Information:

The following calculations estimate the time to failure of cables exposed to a specified hot gas layer.

Parameters in YELLOW CELLS are Entered by the User.

Parameters in GREEN CELLS are Automatically Selected from the DROP DOWN MENU or SELECT CABLE BUTTON.

All subsequent output values are calculated by the spreadsheet and based on values specified in the input parameters. This spreadsheet is protected and secure to avoid errors due to a wrong entry in a cell(s). The chapter in the NUREG should be read before an analysis is made.

Project / Inspection Title:	NUREG-1805 Supplement 1 Example 19.11-5b

INPUT PARAMETERS

Cable Diameter	26.67	mm
Cable Mass per Unit Length	1.38	kg/m
Cable Jacket Thickness	1.524	mm
Ambient Air Temperature	21	°C
Failure Temperature	200	°C
Maximum Time	4000	s
Conduit Thickness	0.00	mm
Conduit Outside Diameter	0.00	mm
Cable Density	2477.41	kg/m³
Cable Insulation Type (Thermoplastic or Thermoset)	Thermoplastic	
Cable Function (Control, Instrumentation or Power)	Control	
Wire Gauge (AWG)	10	
Number of Conductors	19	
Cable Location	Cable Tray	

Do Not Enter Any Values in the Green Boxes!

They are entered automatically based on the cable selection.

Select Cable

Click to Select Source for Exposure Gas Temperature Profile:

Natural Ventilation - Method of McCaffrey, Quintiere, Harkleroad (MQH)
Forced Ventilation - Method of Foote, Pagni, and Alvares (FPA)
Forced Ventilation - Method of Deal and Beyler
Room Fire with Closed Door
Within Fire Plume
User Defined

Selecting one of these items will automatically transfer you to the appropriate spreadsheet to calculate the exposure gas temperature profile

Warning: You MUST Click the Calculate Button Below when Finished Entering or Changing Data!

Calculate

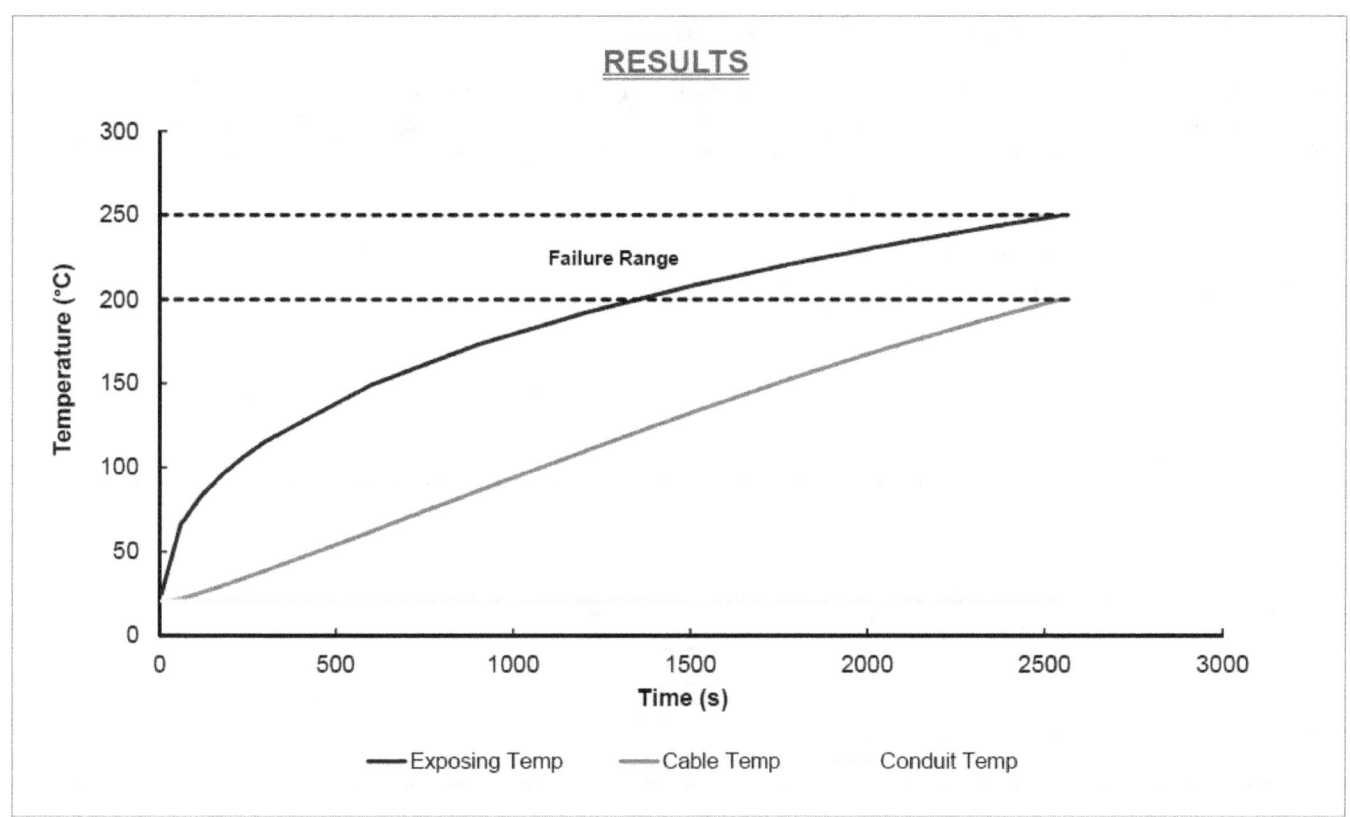

RESULTS

Exposing Temp — Cable Temp — Conduit Temp

EXPOSURE GAS TEMPERATURE PROFILE		
Time (s)	Gas Temperature (°C)	Gas Temperature (°K)
0	21.00	294.15
60	65.98	339.13
120	83.30	356.45
180	96.11	369.26
240	106.61	379.76
300	115.63	388.78
600	149.16	422.31
900	173.02	446.17
1200	192.00	465.15
1500	207.93	481.08
1800	221.72	494.87
2100	233.94	507.09
2400	244.93	518.08
2700	254.93	528.08
3000	264.11	537.26
3300	272.60	545.75
3600	280.52	553.67

Answer: Cable fails at 42.4 minutes

COMPARTMENT WITH THERMALLY THICK/THIN BOUNDARIES

The following calculations estimate the hot gas layer temperature and smoke layer height in enclosure fire.
Parameters in YELLOW CELLS are Entered by the User.
Parameters in GREEN CELLS are Automatically Selected from the DROP DOWN MENU for the Material Selected.
All subsequent output values are calculated by the spreadsheet and based on values specified in the input parameters. This spreadsheet is protected
and secure to avoid errors due to a wrong entry in a cell(s). The chapter in the NUREG should be read before an analysis is made.

Project / Inspection Title:	NUREG-1805 Supplement 1 Example 19.11-5c

INPUT PARAMETERS

COMPARTMENT INFORMATION

Compartment Width (w_c)	5.00	m
Compartment Length (l_c)	5.00	m
Compartment Height (h_c)	3.50	m
Interior Lining Thickness (δ)	30.00	cm

AMBIENT CONDITIONS

Ambient Air Temperature (T_a)	21.00	°C
Specific Heat of Air (c_p)	1.00	kJ/kg-K
Ambient Air Density (ρ_a)	1.20	kg/m^3

Note: Ambient Air Density (ρ_a) will automatically correct with Ambient Air Temperature (T_a) Input

THERMAL PROPERTIES OF COMPARTMENT ENCLOSING SURFACES FOR

Interior Lining Thermal Inertia ($k\rho c$)	2.9	(kW/m^2-K)^2-sec
Interior Lining Thermal Conductivity (k)	0.0016	kW/m-K
Interior Lining Specific Heat (c)	0.75	kJ/kg-K
Interior Lining Density (ρ)	2400	kg/m^3

THERMAL PROPERTIES FOR COMMON INTERIOR LINING MATERIALS

Material	$k\rho c$ $(kW/m^2-K)^2$-sec	k (kW/m-K)	c (kJ/kg-K)	ρ (kg/m^3)	Select Material
					Concrete ▾
Aluminum (pure)	500	0.206	0.895	2710	**Scroll to desired material**
Steel (0.5% Carbon)	197	0.054	0.465	7850	**Click on selection**
Concrete	2.9	0.0016	0.75	2400	
Brick	1.7	0.0008	0.8	2600	
Glass, Plate	1.6	0.00076	0.8	2710	
Brick/Concrete Block	1.2	0.00073	0.84	1900	
Gypsum Board	0.18	0.00017	1.1	960	
Plywood	0.16	0.00012	2.5	540	
F ber Insulation Board	0.16	0.00053	1.25	240	
Chipboard	0.15	0.00015	1.25	800	
Aerated Concrete	0.12	0.00026	0.96	500	
Plasterboard	0.12	0.00016	0.84	950	
Calcium Silicate Board	0.098	0.00013	1.12	700	
Alumina Silicate Block	0.036	0.00014	1	260	
Glass Fiber Insulation	0.0018	0.000037	0.8	60	
Expanded Polystyrene	0.001	0.000034	1.5	20	
User Specified Value	Enter Value	Enter Value	Enter Value	Enter Value	

Reference: Klote, J., J. Milke, Principles of Smoke Management, 2002 Page 270.

COMPARTMENT MASS VENTILATION FLOW RATE

Forced Ventilation Flow Rate (m)   0.47 m^3/s

FIRE SPECIFICATIONS

Fire Heat Release Rate (Q)   500.00 kW

**Click Here to Calculate this
Sheet and Return to THIEF**

CHAPTER 2. PREDICTING HOT GAS LAYER TEMPERATURE
IN A ROOM FIRE
WITH FORCED VENTILATION

Version 1805.1
(SI Units)

Compartment Hot Gas Layer Temperature With Forced Ventilation

$$\Delta T_g/T_a = 0.63(Q/mc_pT_a)^{0.72}(h_kA_T/mc_p)^{-0.36}$$

$$\Delta T_g = T_g - T_a$$

$$T_g = \Delta T_g + T_a$$

Results	Time After Ignition (t)		h_k (kW/m²-K)	$\Delta T_g/T_a$	ΔT_g (°K)	T_g (°K)	T_g (°C)	T_g (°F)
	(min)	(sec)						
	0	0	-	-	-	294.00	21.00	69.80
	1	60	0.22	0.35	102.69	396.69	123.69	254.64
	2	120	0.16	0.40	116.33	410.33	137.33	279.20
	3	180	0.13	0.43	125.14	419.14	146.14	295.05
	4	240	0.11	0.45	131.79	425.79	152.79	307.03
	5	300	0.10	0.47	137.19	431.19	158.19	316.75
	10	600	0.07	0.53	155.42	449.42	176.42	349.56
	15	900	0.06	0.57	167.19	461.19	188.19	370.74
	20	1200	0.05	0.60	176.08	470.08	197.08	386.74
	25	1500	0.04	0.62	183.29	477.29	204.29	399.73
	30	1800	0.04	0.64	189.41	483.41	210.41	410.74
	35	2100	0.04	0.66	194.74	488.74	215.74	420.33
	40	2400	0.03	0.68	199.47	493.47	220.47	428.85
	45	2700	0.03	0.69	203.75	497.75	224.75	436.55
	50	3000	0.03	0.71	207.65	501.65	228.65	443.57
	55	3300	0.03	0.72	211.24	505.24	232.24	450.04
	60	3600	0.03	0.73	214.58	508.58	235.58	456.04

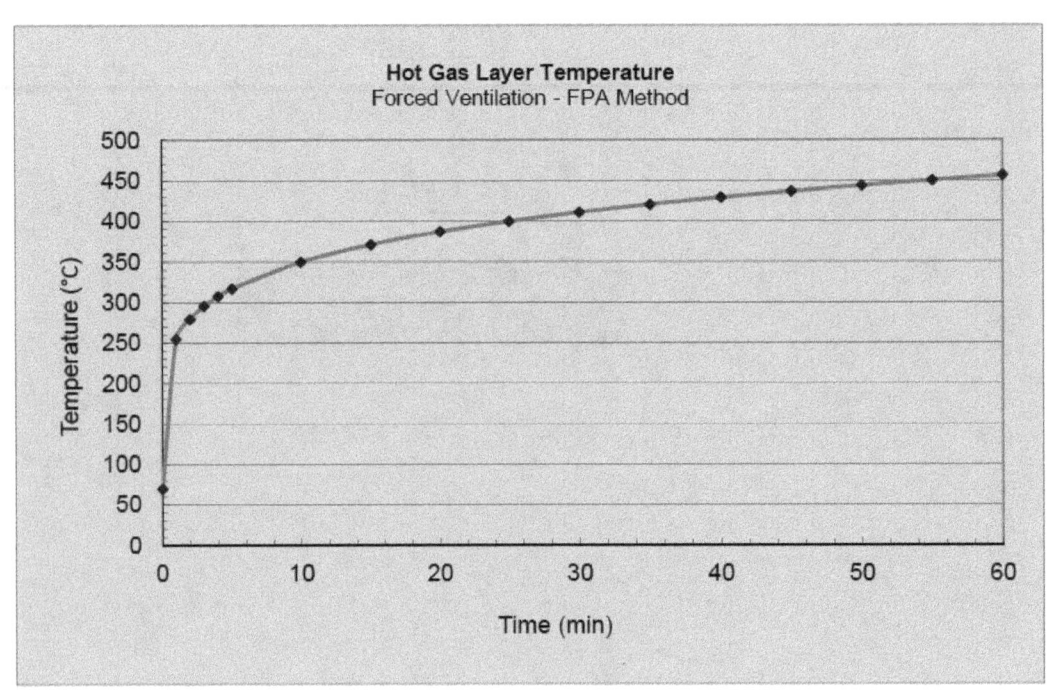

Results	Time After Ignition (t)		h_k	ΔT_g	T_g	T_g	T_g
	(min)	(sec)	(kW/m²-K)	(°K)	(°K)	(°C)	(°F)
	0	0	-	-	294.00	21.00	69.80
	1	60	0.09	44.98	338.98	65.98	150.76
	2	120	0.06	62.30	356.30	83.30	181.93
	3	180	0.05	75.11	369.11	96.11	205.00
	4	240	0.04	85.61	379.61	106.61	223.89
	5	300	0.04	94.63	388.63	115.63	240.14
	10	600	0.03	128.16	422.16	149.16	300.49
	15	900	0.02	152.02	446.02	173.02	343.44
	20	1200	0.02	171.00	465.00	192.00	377.60
	25	1500	0.02	186.93	480.93	207.93	406.27
	30	1800	0.02	200.72	494.72	221.72	431.10
	35	2100	0.01	212.94	506.94	233.94	453.10
	40	2400	0.01	223.93	517.93	244.93	472.87
	45	2700	0.01	233.93	527.93	254.93	490.87
	50	3000	0.01	243.11	537.11	264.11	507.39
	55	3300	0.01	251.60	545.60	272.60	522.69
	60	3600	0.01	259.52	553.52	280.52	536.93

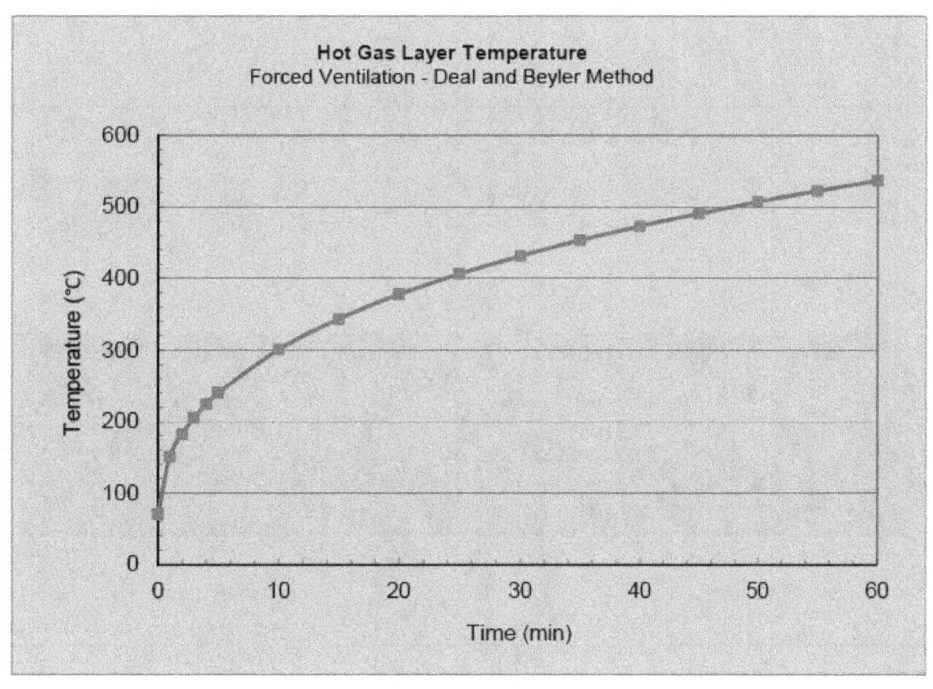

Summary of Results

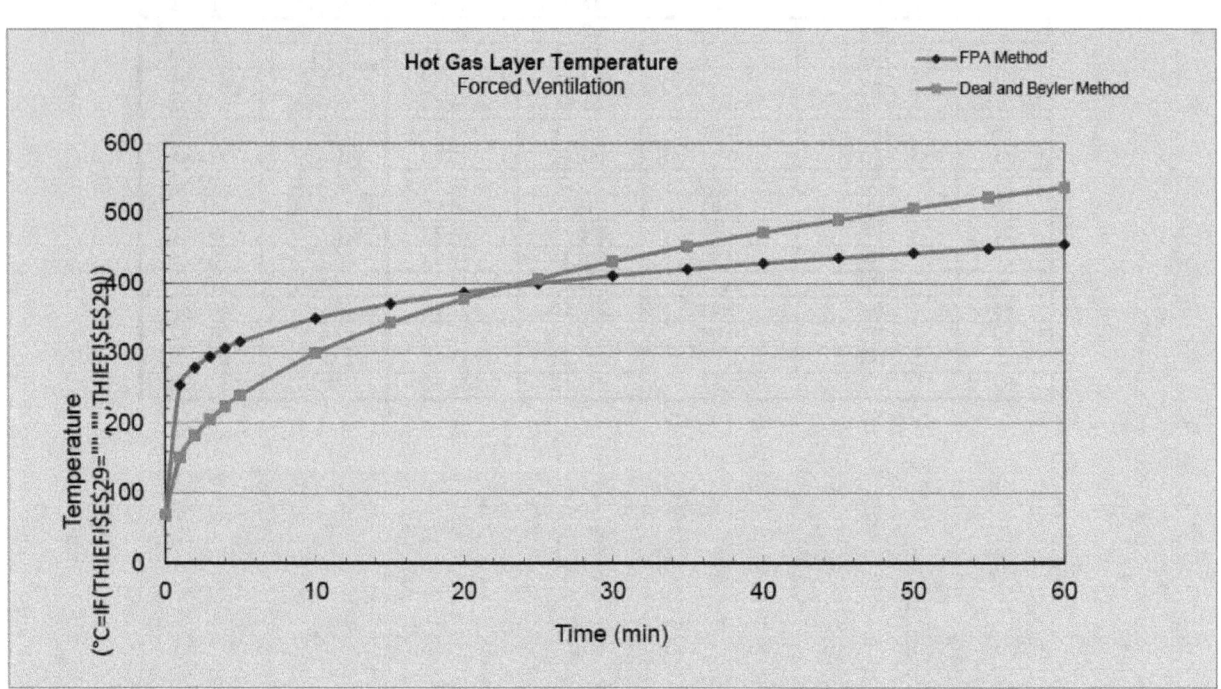

NOTE:
The above calculations are based on principles developed in the SFPE Handbook of Fire Protection Engineering, 2nd Edition, 1995. Calculations are based on certain assumptions and have inherent limitations. The results of such calculations may or may not have reasonable predictive capabilities for a given situation, and should only be interpreted by an informed user. Although each calculation in the spreadsheet has been verified with the results of hand calculation, there is no absolute guarantee of the accuracy of these calculations. Any questions, comments, concerns, and suggestions, or to report an error(s) in the spreadsheets, please send an email to David.Stroup@nrc.gov and Naeem.Iqbal@nrc.gov or MarkHenry.Salley@nrc.gov.

Prepared by: _____   Date: _____   Organization: _____

Checked by: _____   Date: _____   Organization: _____

Additional Information:

The following calculations estimate the time to failure of cables exposed to a specified hot gas layer.

Parameters in YELLOW CELLS are Entered by the User.

Parameters in GREEN CELLS are Automatically Selected from the DROP DOWN MENU or SELECT CABLE BUTTON.

All subsequent output values are calculated by the spreadsheet and based on values specified in the input parameters. This spreadsheet is protected and secure to avoid errors due to a wrong entry in a cell(s). The chapter in the NUREG should be read before an analysis is made.

Project / Inspection Title:	NUREG-1805 Supplement 1 Example 19.11-5c

INPUT PARAMETERS

Cable Diameter	26.67	mm
Cable Mass per Unit Length	1.38	kg/m
Cable Jacket Thickness	1.524	mm
Ambient Air Temperature	21	°C
Failure Temperature	200	°C
Maximum Time	4000	s
Conduit Thickness	4.90	mm
Conduit Outside Diameter	73.00	mm
Cable Density	2477.41	kg/m³
Cable Insulation Type (Thermoplastic or Thermoset)	Thermoplastic	
Cable Function (Control, Instrumentation or Power)	Control	
Wire Gauge (AWG)	10	
Number of Conductors	19	
Cable Location	Conduit - Rigid	

Do Not Enter Any Values in the Green Boxes!

They are entered automatically based on the cable selection.

Select Cable

Click to Select Source for Exposure Gas Temperature Profile:

Natural Ventilation - Method of McCaffrey, Quintiere, Harkleroad (MQH)
Forced Ventilation - Method of Foote, Pagni, and Alvares (FPA)
Forced Ventilation - Method of Deal and Beyler
Room Fire with Closed Door
Within Fire Plume
User Defined

Selecting one of these items will automatically transfer you to the appropriate spreadsheet to calculate the exposure gas temperature profile

Warning: You MUST Click the Calculate Button Below when Finished Entering or Changing Data!

Calculate

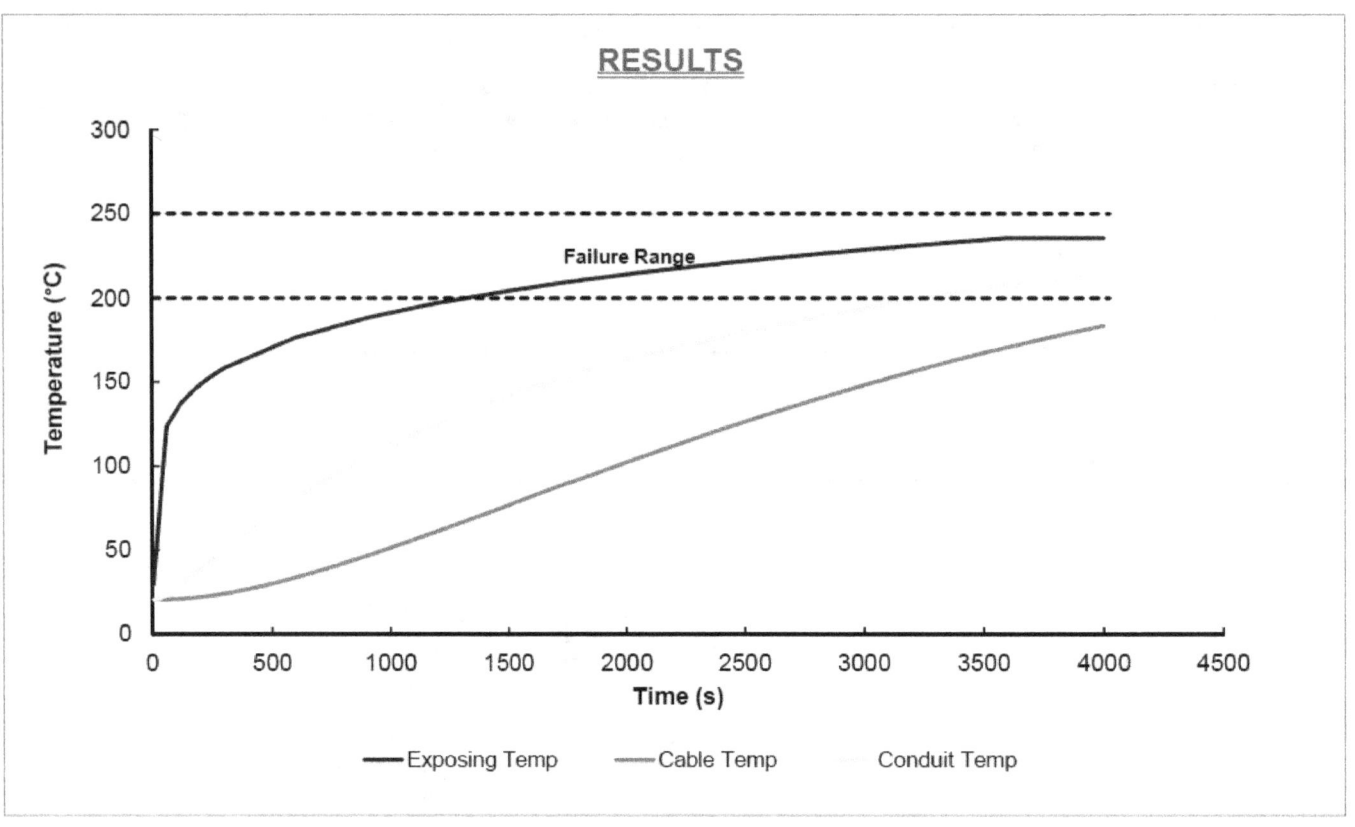

RESULTS

EXPOSURE GAS TEMPERATURE PROFILE		
Time (s)	Gas Temperature (°C)	Gas Temperature (°K)
0	21.00	294.15
60	123.69	396.84
120	137.33	410.48
180	146.14	419.29
240	152.79	425.94
300	158.19	431.34
600	176.42	449.57
900	188.19	461.34
1200	197.08	470.23
1500	204.29	477.44
1800	210.41	483.56
2100	215.74	488.89
2400	220.47	493.62
2700	224.75	497.90
3000	228.65	501.80
3300	232.24	505.39
3600	235.58	508.73

Answer: Cable does not reach failure temperature in 4000.3 seconds

CHAPTER 2. PREDICTING HOT GAS LAYER TEMPERATURE IN A ROOM FIRE WITH FORCED VENTILATION

Version 1805.1
(SI Units)

COMPARTMENT WITH THERMALLY THICK/THIN BOUNDARIES

The following calculations estimate the hot gas layer temperature and smoke layer height in enclosure fire.

Parameters in YELLOW CELLS are Entered by the User.

Parameters in GREEN CELLS are Automatically Selected from the DROP DOWN MENU for the Material Selected.

All subsequent output values are calculated by the spreadsheet and based on values specified in the input parameters. This spreadsheet is protected and secure to avoid errors due to a wrong entry in a cell(s). The chapter in the NUREG should be read before an analysis is made.

Project / Inspection Title:	NUREG-1805 Supplement 1 Example 19.11-5d

INPUT PARAMETERS

COMPARTMENT INFORMATION

Compartment Width (w_c)	5.00	m
Compartment Length (l_c)	5.00	m
Compartment Height (h_c)	3.50	m
Interior Lining Thickness (δ)	30.00	cm

AMBIENT CONDITIONS

Ambient Air Temperature (T_a)	21.00	°C
Specific Heat of Air (c_p)	1.00	kJ/kg-K
Ambient Air Density (ρ_a)	1.20	kg/m^3

Note: Ambient Air Density (ρ_a) will automatically correct with Ambient Air Temperature (T_a) Input

THERMAL PROPERTIES OF COMPARTMENT ENCLOSING SURFACES FOR

Interior Lining Thermal Inertia ($k\rho c$)	2.9	(kW/m^2-K)^2-sec
Interior Lining Thermal Conductivity (k)	0.0016	kW/m-K
Interior Lining Specific Heat (c)	0.75	kJ/kg-K
Interior Lining Density (ρ)	2400	kg/m^3

THERMAL PROPERTIES FOR COMMON INTERIOR LINING MATERIALS

Material	kρc (kW/m²-K)²-sec	k (kW/m-K)	c (kJ/kg-K)	ρ (kg/m³)	Select Material
					Concrete ▾
Aluminum (pure)	500	0.206	0.895	2710	**Scroll to desired material**
Steel (0.5% Carbon)	197	0.054	0.465	7850	**Click on selection**
Concrete	2.9	0.0016	0.75	2400	
Brick	1.7	0.0008	0.8	2600	
Glass, Plate	1.6	0.00076	0.8	2710	
Brick/Concrete Block	1.2	0.00073	0.84	1900	
Gypsum Board	0.18	0.00017	1.1	960	
Plywood	0.16	0.00012	2.5	540	
F ber Insulation Board	0.16	0.00053	1.25	240	
Chipboard	0.15	0.00015	1.25	800	
Aerated Concrete	0.12	0.00026	0.96	500	
Plasterboard	0.12	0.00016	0.84	950	
Calcium Silicate Board	0.098	0.00013	1.12	700	
Alumina Silicate Block	0.036	0.00014	1	260	
Glass Fiber Insulation	0.0018	0.000037	0.8	60	
Expanded Polystyrene	0.001	0.000034	1.5	20	
User Specified Value	Enter Value	Enter Value	Enter Value	Enter Value	

Reference: Klote, J., J. Milke, Principles of Smoke Management, 2002 Page 270.

COMPARTMENT MASS VENTILATION FLOW RATE

Forced Ventilation Flow Rate (m)   0.47 m³/s

FIRE SPECIFICATIONS

Fire Heat Release Rate (Q)   500.00 kW

**Click Here to Calculate this
Sheet and Return to THIEF**

CHAPTER 2. PREDICTING HOT GAS LAYER TEMPERATURE IN A ROOM FIRE WITH FORCED VENTILATION

Version 1805.1
(SI Units)

Compartment Hot Gas Layer Temperature With Forced Ventilation

$$\Delta T_g/T_a = 0.63(Q/mc_pT_a)^{0.72}(h_kA_T/mc_p)^{-0.36}$$

$$\Delta T_g = T_g - T_a$$

$$T_g = \Delta T_g + T_a$$

	Time After Ignition (t)		h_k	$\Delta T_g/T_a$	ΔT_g	T_g	T_g	T_g
Results	(min)	(sec)	(kW/m²-K)		(°K)	(°K)	(°C)	(°F)
	0	0	-	-	-	294.00	21.00	69.80
	1	60	0.22	0.35	102.69	396.69	123.69	254.64
	2	120	0.16	0.40	116.33	410.33	137.33	279.20
	3	180	0.13	0.43	125.14	419.14	146.14	295.05
	4	240	0.11	0.45	131.79	425.79	152.79	307.03
	5	300	0.10	0.47	137.19	431.19	158.19	316.75
	10	600	0.07	0.53	155.42	449.42	176.42	349.56
	15	900	0.06	0.57	167.19	461.19	188.19	370.74
	20	1200	0.05	0.60	176.08	470.08	197.08	386.74
	25	1500	0.04	0.62	183.29	477.29	204.29	399.73
	30	1800	0.04	0.64	189.41	483.41	210.41	410.74
	35	2100	0.04	0.66	194.74	488.74	215.74	420.33
	40	2400	0.03	0.68	199.47	493.47	220.47	428.85
	45	2700	0.03	0.69	203.75	497.75	224.75	436.55
	50	3000	0.03	0.71	207.65	501.65	228.65	443.57
	55	3300	0.03	0.72	211.24	505.24	232.24	450.04
	60	3600	0.03	0.73	214.58	508.58	235.58	456.04

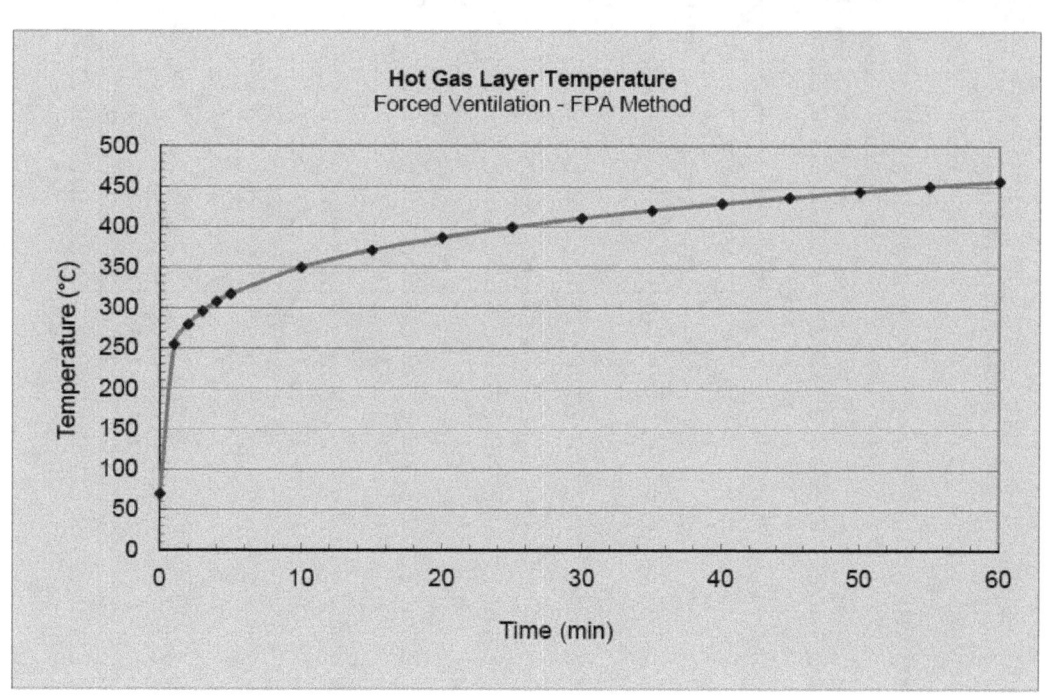

Hot Gas Layer Temperature
Forced Ventilation - FPA Method

19-103

Results	Time After Ignition (t)		h_k	ΔT_g	T_g	T_g	T_g
	(min)	(sec)	(kW/m²-K)	(°K)	(°K)	(°C)	(°F)
	0	0	-	-	294.00	21.00	69.80
	1	60	0.09	44.98	338.98	65.98	150.76
	2	120	0.06	62.30	356.30	83.30	181.93
	3	180	0.05	75.11	369.11	96.11	205.00
	4	240	0.04	85.61	379.61	106.61	223.89
	5	300	0.04	94.63	388.63	115.63	240.14
	10	600	0.03	128.16	422.16	149.16	300.49
	15	900	0.02	152.02	446.02	173.02	343.44
	20	1200	0.02	171.00	465.00	192.00	377.60
	25	1500	0.02	186.93	480.93	207.93	406.27
	30	1800	0.02	200.72	494.72	221.72	431.10
	35	2100	0.01	212.94	506.94	233.94	453.10
	40	2400	0.01	223.93	517.93	244.93	472.87
	45	2700	0.01	233.93	527.93	254.93	490.87
	50	3000	0.01	243.11	537.11	264.11	507.39
	55	3300	0.01	251.60	545.60	272.60	522.69
	60	3600	0.01	259.52	553.52	280.52	536.93

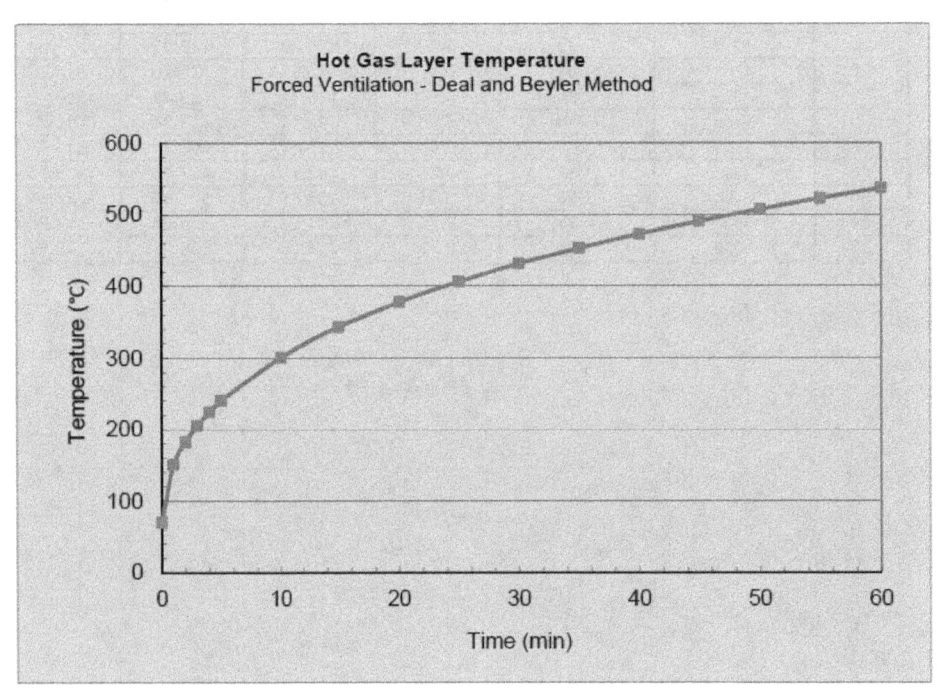

Summary of Results

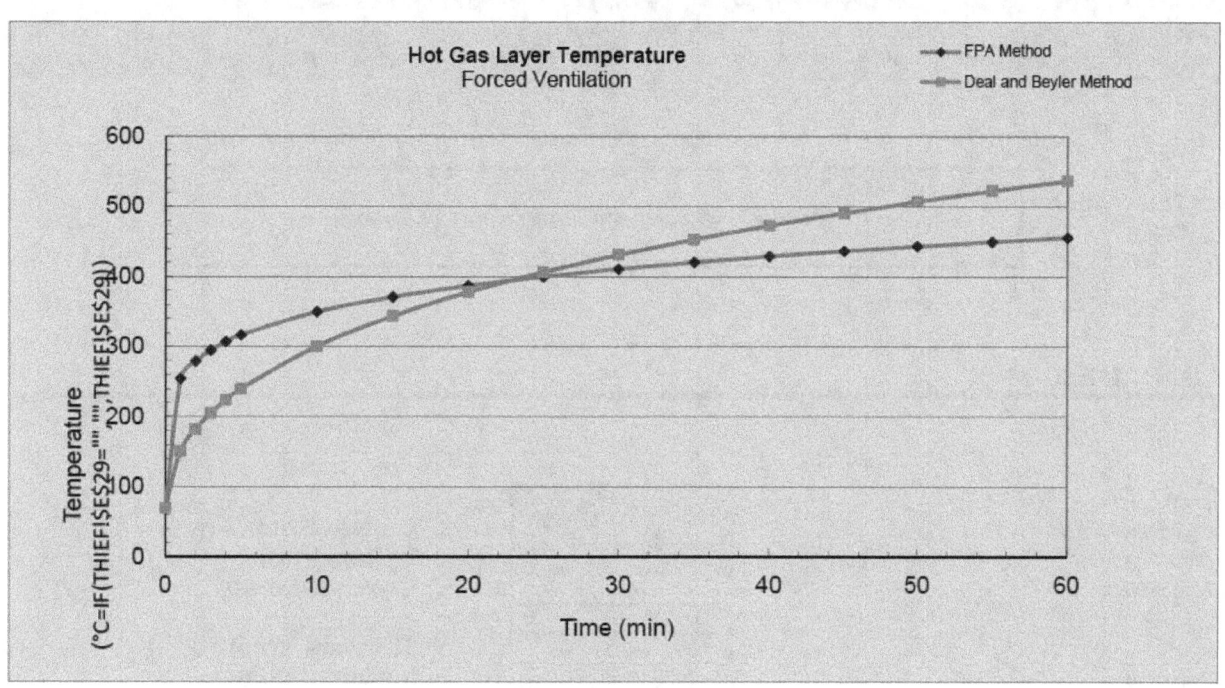

NOTE:

The above calculations are based on principles developed in the SFPE Handbook of Fire Protection Engineering, 2nd Edition, 1995. Calculations are based on certain assumptions and have inherent limitations. The results of such calculations may or may not have reasonable predictive capabilities for a given situation, and should only be interpreted by an informed user. Although each calculation in the spreadsheet has been verified with the results of hand calculation, there is no absolute guarantee of the accuracy of these calculations. Any questions, comments, concerns, and suggestions, or to report an error(s) in the spreadsheets, please send an email to David.Stroup@nrc.gov and Naeem.Iqbal@nrc.gov or MarkHenry.Salley@nrc.gov.

Prepared by: [ ] Date: [ ] Organization: [ ]

Checked by: [ ] Date: [ ] Organization: [ ]

Additional Information:

The following calculations estimate the time to failure of cables exposed to a specified hot gas layer.

Parameters in YELLOW CELLS are Entered by the User.

Parameters in GREEN CELLS are Automatically Selected from the DROP DOWN MENU or SELECT CABLE BUTTON.

All subsequent output values are calculated by the spreadsheet and based on values specified in the input parameters. This spreadsheet is protected and secure to avoid errors due to a wrong entry in a cell(s). The chapter in the NUREG should be read before an analysis is made.

Project / Inspection Title:	NUREG-1805 Supplement 1 Example 19.11-5d

INPUT PARAMETERS

Cable Diameter	26.67	mm
Cable Mass per Unit Length	1.38	kg/m
Cable Jacket Thickness	1.524	mm
Ambient Air Temperature	21	°C
Failure Temperature	200	°C
Maximum Time	4000	s
Conduit Thickness	4.90	mm
Conduit Outside Diameter	73.00	mm
Cable Density	2477.41	kg/m³
Cable Insulation Type (Thermoplastic or Thermoset)	Thermoplastic	
Cable Function (Control, Instrumentation or Power)	Control	
Wire Gauge (AWG)	10	
Number of Conductors	19	
Cable Location	Conduit - Rigid	

Do Not Enter Any Values in the Green Boxes!

They are entered automatically based on the cable selection.

Select Cable

Click to Select Source for Exposure Gas Temperature Profile:

Natural Ventilation - Method of McCaffrey, Quintiere, Harkleroad (MQH)
Forced Ventilation - Method of Foote, Pagni, and Alvares (FPA)
Forced Ventilation - Method of Deal and Beyler
Room Fire with Closed Door
Within Fire Plume
User Defined

Selecting one of these items will automatically transfer you to the appropriate spreadsheet to calculate the exposure gas temperature profile

Warning: You MUST Click the Calculate Button Below when Finished Entering or Changing Data!

Calculate

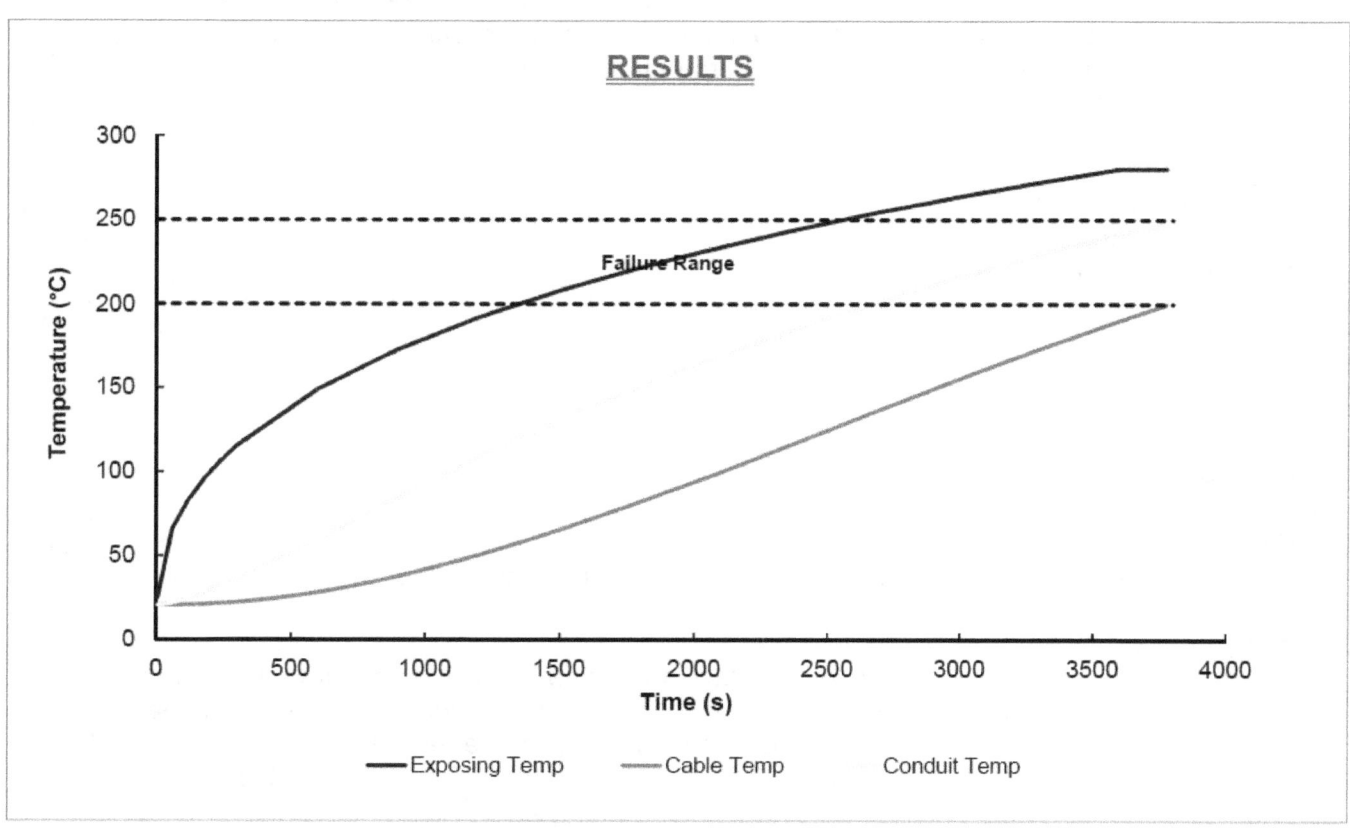

EXPOSURE GAS TEMPERATURE PROFILE		
Time (s)	Gas Temperature (°C)	Gas Temperature (°K)
0	21.00	294.15
60	65.98	339.13
120	83.30	356.45
180	96.11	369.26
240	106.61	379.76
300	115.63	388.78
600	149.16	422.31
900	173.02	446.17
1200	192.00	465.15
1500	207.93	481.08
1800	221.72	494.87
2100	233.94	507.09
2400	244.93	518.08
2700	254.93	528.08
3000	264.11	537.26
3300	272.60	545.75
3600	280.52	553.67

Answer: Cable fails at 63 minutes

Example Problem 19.11-6 (SI Units)
(Cable Within Plume)

Problem Statement
Consider a concrete compartment that is 5.00 m (16.40 ft) wide x 5.00 m (16.40 ft) long x 3.5 m (11.48 ft) high (w_c x l_c x h_c). A cable tray is located 1.83 m (6 ft) above a burning electrical cabinet. The fire is constant with a HRR of 250kW and a combustible area of 0.28 m^2 (3 ft^2). Compute the time to cable failure using a thermoplastic cable located in the cable tray with the following parameters:

- Control
- 14 AWG
- 9 conductors
- PE insulation
- PVC jacket
- Dekoron
- Model : 1735

Solution

Purpose:

      (1)   Determine the time to failure of the cable.

Assumptions:

      (1)   The cable is surrounded by a uniform temperature fire plume.
      (2)   The cable properties do not change with temperature increase.
      (3)   Failure is indicated when the temperature inside the cable jacket exceeds 200 °C (392 °F) for the thermoplastic cable.

Spreadsheet (FDT^s) Information:

      Use the following FDT^s:
          (a)   09_Plume_Temperature_Calculations_Sup1_SI.xls
          (b)   19_THIEF_Thermally_Induced_Electrical_Failure_of_Cables_Sup1_SI.xls (click on *THIEF*)

FDT^s Input Parameters
      -Gas temperature around the cable as a function time:
      Use 09_Plume_Tempearture_Calculations_Sup1_SI.xls and the parameters presented above to determine the plume centerline temperature
      On the THIEF spreadsheet, select: Plume_Calculations
      -Press "Select Cable" Button
      -Choose appropriate cable from list
          Control – 14 AWG – 9 conductor – Dekoron – 1735 – PE – PVC
      -Choose appropriate Cable Location ("Cable Tray")
      -Press "Calculate" Button

Results*

Thermoplastic cable failure time (minutes)	3.9

*see spreadsheets

The results show a thermoplastic cable failing in 3.9 minutes.

The following calculations estimate the centerline plume temperature in a compartment fire.
Parameters should be specified ONLY IN THE YELLOW INPUT PARAMETER BOXES.
All subsequent output values are calculated by the spreadsheet and based on values specified in the input parameters. This spreadsheet is protected and secure to avoid errors due to a wrong entry in a cell(s). The chapter in the NUREG should be read before an analysis is made.

Project / Inspection Title:	NUREG-1805 Supplement 1 Example 19.11-6

INPUT PARAMETERS

Heat Release Rate of the Fire (Q)	250.00	kW
Elevation Above the Fire Source (z)	1.83	m
Area of Combustble Fuel (A_c)	0.28	m^2
Ambient Air Temperature (T_a)	21.00	°C

AMBIENT CONDITIONS

Specific Heat of Air (c_p)	1.00	kJ/kg-K
Ambient Air Density (ρ_a)	1.20	kg/m^3
Acceleration of Gravity (g)	9.81	m/sec^2
Convective Heat Release Fraction (χ_c)	0.70	

NOTE: Ambient Air Density (ρ_a) will automatically correct with Ambient Air Temperature (T_a) Input

Click Here to Calculate this Sheet and Return to THIEF

Mean Flame Height Calculation

$$L = -1.02D + 0.235 (Q^{2/5})$$

Where,
   L = mean flame height (m)
   Q = heat release rate of fire (kW)
   D = fire diameter (m)

$L =$ **1.53 m**

Centerline Plume Temperature Calculation

$$T_{p(centerline)} - T_a = 9.1 (T_a/g\, c_p^2\, \rho_a^2)^{1/3} Q_c^{2/3} (z - z_0)^{-5/3}$$

$T_{p(centerline)} - T_a =$ **328.59 °K**

$T_{p(centerline)} =$ **622.59 °K**

Answer	$T_{p(centerline)} =$	349.59 °C	661.27 °F

NOTE:

The above calculations are based on principles developed in the SFPE Handbook of Fire Protection Engineering, 2nd Edition, 1995. Calculations are based on certain assumptions and have inherent limitations. The results of such calculations may or may not have reasonable predictive capabilities for a given situation, and should only be interpreted by an informed user. Although each calculation in the spreadsheet has been verified with the results of hand calculation, there is no absolute guarantee of the accuracy of these calculations. Any questions, comments, concerns, and suggestions, or to report an error(s) in the spreadsheets, please send an email to David.Stroup@nrc.gov and Naeem.Iqbal@nrc.gov or MarkHenry.Salley@nrc.gov.

Prepared by: [   ]   Date: [   ]   Organization: [   ]

Checked by: [   ]   Date: [   ]   Organization: [   ]

Additional Information:

The following calculations estimate the time to failure of cables exposed to a specified hot gas layer.

Parameters in YELLOW CELLS are Entered by the User.

Parameters in GREEN CELLS are Automatically Selected from the DROP DOWN MENU or SELECT CABLE BUTTON.

All subsequent output values are calculated by the spreadsheet and based on values specified in the input parameters. This spreadsheet is protected and secure to avoid errors due to a wrong entry in a cell(s). The chapter in the NUREG should be read before an analysis is made.

Project / Inspection Title:	NUREG-1805 Supplement 1 Example 19.11-6

INPUT PARAMETERS

Cable Diameter	16.51	mm
Cable Mass per Unit Length	0.37	kg/m
Cable Jacket Thickness	1.524	mm
Ambient Air Temperature	21	°C
Failure Temperature	200	°C
Maximum Time	4000	s
Conduit Thickness	0.00	mm
Conduit Outside Diameter	0.00	mm
Cable Density	1723.93	kg/m³
Cable Insulation Type (Thermoplastic or Thermoset)	Thermoplastic	
Cable Function (Control, Instrumentation or Power)	Control	
Wire Gauge (AWG)	14	
Number of Conductors	9	
Cable Location	Cable Tray	

Do Not Enter Any Values in the Green Boxes!

They are entered automatically based on the cable selection.

Select Cable

Click to Select Source for Exposure Gas Temperature Profile:

Natural Ventilation - Method of McCaffrey, Quintiere, Harkleroad (MQH)
Forced Ventilation - Method of Foote, Pagni, and Alvares (FPA)
Forced Ventilation - Method of Deal and Beyler
Room Fire with Closed Door
Within Fire Plume
User Defined

Selecting one of these items will automatically transfer you to the appropriate spreadsheet to calculate the exposure gas temperature profile

Warning: You MUST Click the Calculate Button Below when Finished Entering or Changing Data!

Calculate

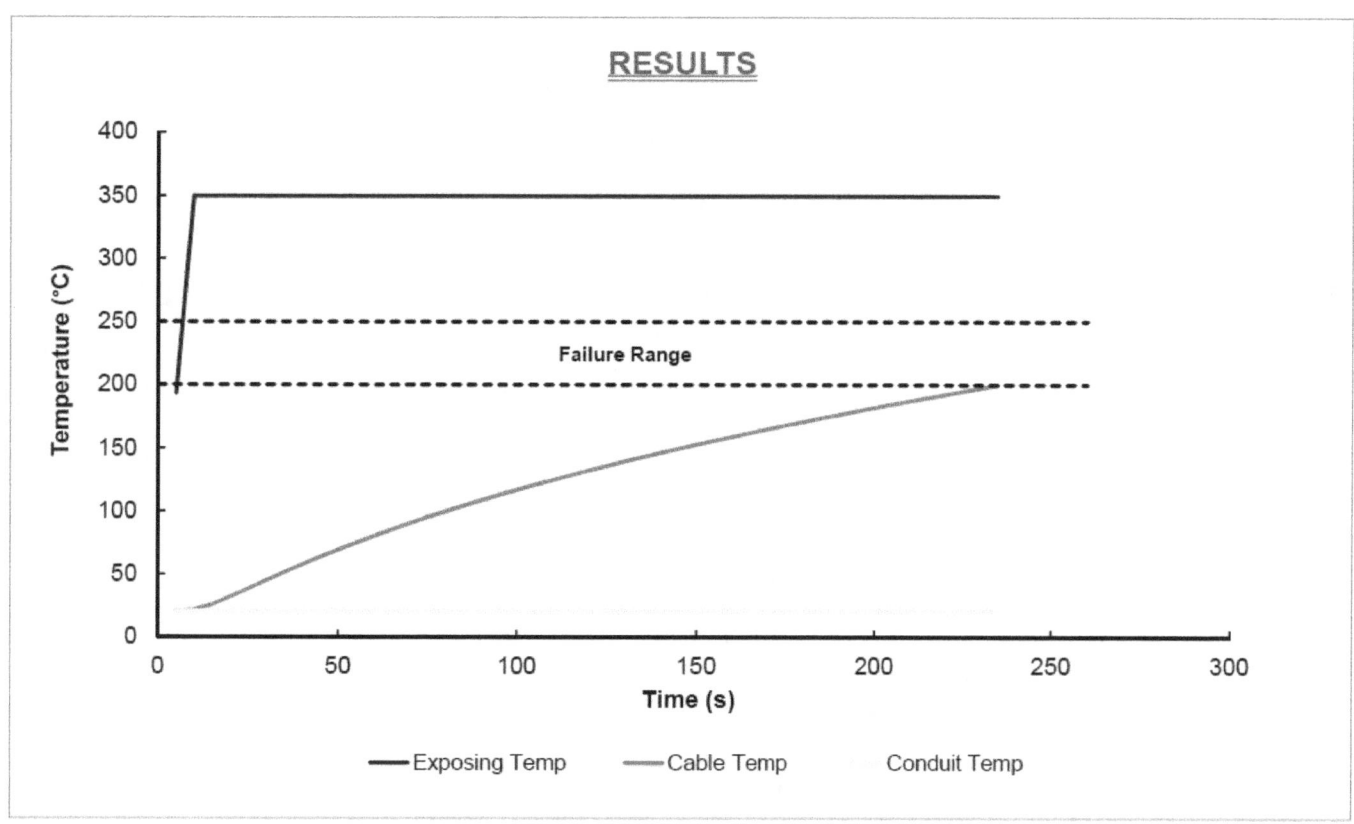

RESULTS

EXPOSURE GAS TEMPERATURE PROFILE		
Time (s)	Gas Temperature (°C)	Gas Temperature (°K)
0	21.00	294.15
10	349.59	622.74
2000	349.59	622.74
4000	349.59	622.74

Answer: Cable fails at 3.9 minutes

NRC FORM 335
(12-2010)
NRCMD 3.7

U.S. NUCLEAR REGULATORY COMMISSION

BIBLIOGRAPHIC DATA SHEET

(See instructions on the reverse)

1. REPORT NUMBER
(Assigned by NRC, Add Vol., Supp., Rev., and Addendum Numbers, if any.)

NUREG-1805
Supplement 1
Volume 1

2. TITLE AND SUBTITLE

Fire Dynamics Tools (FDTs) - Quantitative Fire Hazard Analysis Methods for the U.S. Nuclear Regulatory Commission Fire Protection Inspection Program - Supplement 1

3. DATE REPORT PUBLISHED

MONTH	YEAR
July	2013

4. FIN OR GRANT NUMBER

5. AUTHOR(S)

D. Stroup, G. Taylor, G. Hausman

6. TYPE OF REPORT

Technical

7. PERIOD COVERED (Inclusive Dates)

8. PERFORMING ORGANIZATION - NAME AND ADDRESS (If NRC, provide Division, Office or Region. U. S. Nuclear Regulatory Commission, and mailing address; if contractor, provide name and mailing address.)

Division of Risk Analysis
Office of Nuclear Regulatory Research
U.S. Nuclear Regulatory Commission
Washington, DC 20555-0001

9. SPONSORING ORGANIZATION - NAME AND ADDRESS (If NRC, type "Same as above", if contractor, provide NRC Division, Office or Region, U. S. Nuclear Regulatory Commission, and mailing address.)

Same as above

10. SUPPLEMENTARY NOTES
M.H. Salley, NRC Project Manager

11. ABSTRACT (200 words or less)
The U.S. Nuclear Regulatory Commission (NRC) has developed quantitative methods, known as "Fire Dynamics Tools" (FDTs), for analyzing the impact of fire and fire protection systems in nuclear power plants (NPPs). These methods have been implemented in spreadsheets and taught at the NRC's quarterly regional inspector workshops. The FDTs were developed using state-of-the-art fire dynamics equations and correlations that were preprogrammed and locked into Microsoft Excel® spreadsheets. These FDTs enable inspectors to perform quick, easy, first-order calculations for potential fire scenarios using today's state-of-the-art principles of fire dynamics. Each FDTs spreadsheet also contains a list of the physical and thermal properties of the materials commonly encountered in NPPs. This NUREG-series report documents a new spreadsheet that has been added to the FDTs suite and describes updates, corrections, and improvements to the existing spreadsheets. The majority of the original FDTs were developed using principles and information from the Society of Fire Protection Engineers (SFPE) Handbook of Fire Protection Engineering, the National Fire Protection Association (NFPA) Fire Protection Handbook, and other fire science literature. The new spreadsheet predicts the behavior of power cables, instrument cables, and control cables during a fire. The thermally-induced electrical failure (THIEF) model was developed by the National Institute of Standards and Technology (NIST) as part of the Cable Response to Live Fire (CAROLFIRE) program sponsored by the NRC. The experiments for CAROLFIRE were conducted at Sandia National Laboratories, Albuquerque, New Mexico. THIEF model predictions have been compared to experimental measurements of instrumented cables in a variety of configurations, and the results indicate that the model is an appropriate analysis tool for NPP applications. The accuracy and simplicity of the THIEF model have been shown to be comparable to that of the activation algorithms for various fire protection devices (e.g., sprinklers, heat and smoke detectors).

12. KEY WORDS/DESCRIPTORS (List words or phrases that will assist researchers in locating the report.)

Fire Dynamics, Fire Hazard Analysis (FHA), Risk-Informed Regulation, Fire Safety, Fire Protection, Probabilistic Risk Assessment (PRA), Fire Modeling, Significance Determination Process

13. AVAILABILITY STATEMENT
unlimited

14. SECURITY CLASSIFICATION

(This Page)
unclassified

(This Report)
unclassified

15. NUMBER OF PAGES

16. PRICE

NUREG-1805
Supplement 1, Vol. 1

Fire Dynamics Tools FDTˢ Quantitative Fire Hazard Analysis Methods for the
U.S. Nuclear Regulatory Commission Fire Protection Inspection Program

July 2013